De Gruyter Graduate

Agarwal, Burgard, Greiner, Wendorff • Electrospinning

Also of interest

Structure Analysis of Advanced Nanomaterials: Nanoworld by High-Resolution Electron Microscopy
Oku, 2014
ISBN 978-3-11-030472-5, e-ISBN 978-3-11-030501-2

Polymer Surface Characterization
Sabbatini (Ed.), 2014
ISBN 978-3-11-027508-7, e-ISBN 978-3-11-028811-7

Nanodispersions
Tadros, 2015
ISBN 978-3-11-029033-2, e-ISBN 978-3-11-029034-9

Biomaterials: Biological Production of Fuels and Chemicals
Luque, Xu (Eds.), 2016
ISBN 978-3-11-033671-9, e-ISBN 978-3-11-033672-6

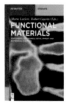

Functional Materials: For Energy, Sustainable Development and Biomedical Sciences
Leclerc, Gauvin (Eds.), 2014
ISBN 978-3-11-030781-8, e-ISBN 978-3-11-030782-5

e-Polymers
Seema Agarwal, Andreas Greiner (Editors-in-Chief)
ISSN 2197-4586, e-ISSN 1618-7229

Seema Agarwal, Matthias Burgard,
Andreas Greiner, Joachim H. Wendorff

Electrospinning

A Practical Guide to Nanofibers

DE GRUYTER

Authors

Prof. Seema Agarwal
University of Bayreuth
Macromolecular Chemistry II
Universitätstr. 30
95447 Bayreuth, Germany

Prof. Andreas Greiner
University of Bayreuth
Macromolecular Chemistry II
Universitätstr. 30
95447 Bayreuth, Germany

Matthias Burgard
University of Bayreuth
Macromolecular Chemistry II
Universitätstr. 30
95447 Bayreuth, Germany

Prof. Joachim H. Wendorff
Philipps-University Marburg
Department of Chemistry
Hans-Meerwein-Str. 6
35032 Marburg, Germany

ISBN 978-3-11-033180-6
e-ISBN (PDF) 978-3-11-033351-0
e-ISBN (EPUB) 978-3-11-038260-0

Library of Congress Cataloging-in-Publication Data
A CIP catalog record for this book has been applied for at the Library of Congress.

Bibliographic information published by the Deutsche Nationalbibliothek
The Deutsche Nationalbibliothek lists this publication in the Deutsche Nationalbibliografie; detailed bibliographic data are available on the Internet at http://dnb.dnb.de.

© 2016 Walter de Gruyter GmbH, Berlin/Boston
Typesetting: Lumina Datamatics
Printing and binding: Hubert & Co. GmbH & Co. KG, Göttingen
♾ Printed on acid-free paper
Printed in Germany

www.degruyter.com

Preface

Electrospinning has matured to the state-of-the-art technique for the preparation of submicrometer-fibers, which is used in technical applications and is also a constant source for novel scientific developments and challenges. It motivates interdisciplinary research and therefore has become a crossover technology. Constantly, novel research wants to utilize the benefits and chances offered by electrospinning. Therefore, it is time for a bench- and desktop reference for those who want to get started in electrospinning, want to solve particular problems, want to probe electrospinning in technical applications or want to use it in teaching with practical manuals for undergraduate as well as graduate students. The readers of the book should feel motivated to impose their problems on what is provided by the book or just feel motivated to try with their own materials something new, taking the information given in the book as a starting point. Although electrospinning looks simple, it is not the case. Numerous parameters have an impact on the final result and complicate reproducibility and understanding of the outcome of experiments. This can be a major obstacle for beginners or technical operators in electrospinning. The successful experiment is a combination of excellent equipment for electrospinning, well-adjusted electrospinning formulation and the application of the right characterization techniques to exploit as much information from the experiment and to gain maximum understanding. The chapters of the book should enable the readers of the book to do electrospinning successfully, enjoy the experiments and discover the undiscovered matter.

The first chapter of the book provides some general information about fibers and their preparations and applications. The second chapter focuses on nanofibers including fibers made by electrospinning. In the third chapter the reader gets information about some selected methods for the characterization of spinning solutions and electrospun fibers. The fourth chapter provides probed procedures for electrospinning under different aspects of processing various polymers, using different kinds of electrospinning formulations, and making complex fiber architectures. Finally, in the fifth chapter selected applications of electrospun fibers are introduced together with the chemistry of the corresponding polymers. Each chapter provides references for further reading and ends by a section "Interesting to know," which should give some further information and should inspire to dive deeper into the subject.

The authors would like to thank numerous coworkers and colleagues who provided the basis for electrospinning, which is condensed in this book.

Contents

1 Introduction and fiber-processing methods

1.1 Fibers

This textbook focuses on a practical approach toward electrospinning, a very unique production technology for fibers with diameters down to the nanometers scale, i.e., to the 10^{-9} m scale, which are composed of natural and synthetic, organic, and inorganic materials or polymers. Fibers are essentially linear, one-dimensional objects with a longitudinal extension. In other words, the length is much larger than the diameter. The aspect ratio, which is defined as the ratio of the longitudinal versus the lateral dimension, can be of the order of just 10 corresponding to rod-type structural elements, but also of the order of 1,000, 10,000 and above, corresponding to wire-type structural elements (Figure 1.1).

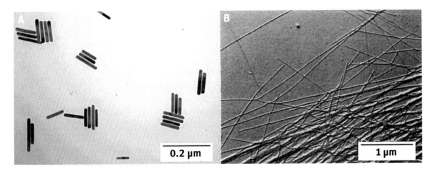

Figure 1.1: Fibers with small (rod type) and large (wire type) aspect ratios: gold nanorods (A) and poly-tetrafluoroethylene nanowires with an aspect ratio of > 1,000 (B). [Figure B is reprinted by permission from Macmillan Publishers Ltd: *Nature*, **1998**, *333*, 55, copyright Nature publishing group.]

Fibers can consist of just one type of material, which is homogeneously distributed throughout the cross section and along the total length of the fibers, thus at all locations having the same internal structure. Yet the composition of fibers can be much more complex in terms of materials, i.e., one material or mixture of materials; internal structure, i.e., amorphous or crystalline; the surface structure, i.e., smooth or porous; and the shape of the cross section, i.e., circular or triangular.

Fibers are ubiquitous in nature as well as in everyday applications. In nature fibers are present not only in spider nets but also in animal skins or as structural elements in plants. In technical applications as well, fibers are key structural elements in different areas, such as textiles, ropes, filters, reinforcement purposes, sensors, light guidance systems, insulation purposes. Many of these applications make use of fibrous nonwovens, which are sheetlike structures with a planar arrangement of fibers (Figure 1.2).

Figure 1.2: **SEM images of nonwovens based on fibers with isotropic fiber orientation and parallel fiber orientation in the plane of the nonwoven.**

The applications introduced here traditionally rely on fibers with a natural or synthetic origin, with the range of diameters from several micrometers to tens of micrometers. Electrospinning allows the reduction of such fiber diameters by several orders of magnitudes, thus opening potential novel areas of applications. To bring the currently developing knowledge on electrospinning, nanofiber preparation and nanofiber properties in line with the well-established knowledge on traditional fiber-making approaches and application activities addressing micrometer-sized fibers, it is instructive to learn these conventional activities in some detail.

1.2 Natural fibers – fibers provided by nature

Fibers are produced by nature in large amounts. Fibrous materials from plants and animals have been the fundament for textile applications for many years [1, 2]. In fact, fibers are in use since many thousands of years. In preindustrial times fibers were mainly of natural origin. For example, they were obtained from plants such as flax, cotton and jute, from animals such as sheep and goats yielding wool, or from insects yielding silk fibers. Natural fibers tend to be classified according to their origin, i.e., vegetable/cellulose, animal/protein, and mineral fibers.

Flax and cotton are common examples of vegetable/cellulose fibers and belong to the oldest fiber types in the world. The common applications of flax fibers are linen fabrics, banknotes, cigarettes, and tea bags. Cotton fibers are soft, fluffy, and short (fiber length: few centimeters), with diameters in the 10 μm range. They can be spun into a yarn called staple fibers. Staple fibers are (generally) spun into long yarns, suitable for subsequent textile processing. Sisal and jute are other common naturally occurring fibers, used for making ropes, twines, paper, cloth, carpets, dartboards, and reinforcements of polymeric matrices. Natural fibers consist of polysaccharides, such as cellulose, hemicellulose, pectin, and polymer lignin in different ratios, depending on the type of fiber and harvesting conditions used for the growth. The fiber structure is complex in which the amorphous lignin binds helically arranged cellulose micro fibrils (diameter of about 10–30 nm), whereas hemicellulose compatibilizes the cellulose and the lignin (Figure 1.3). The mechanical strength of fibers comes from the cellulose microfibrils, which are made up of 30–100 cellulose molecules (each cellulose macromolecule consists of β-D-glucose repeat units). Cellulose is a well-known filler for reinforcement purposes. The details about cellulose-based composites can be found in review articles [3].

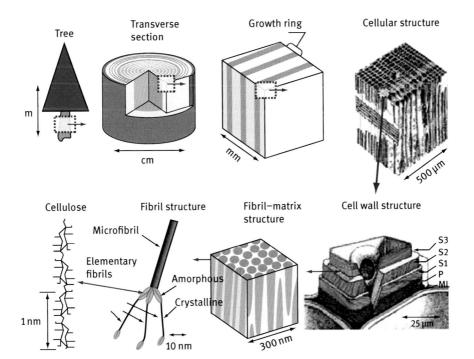

Figure 1.3: Hierarchical structure of wood down to the molecular area. [Reprinted with permission from *Measure. Sci. Technol.* **2011,** *22,* 024005.]

Wool is an elastic fiber, which is in use for textile applications. It grows in staples and is crimped. *Crimp* is a term used to describe the waviness of a fiber. Crimp in a fiber is desirable because it improves the spinning quality of staple fibers. Furthermore, this makes it easy to process the fiber into a variety of forms, such as a spun yarn. For this reason, efforts are undertaken to induce crimp also in synthetic fibers, including nanofibers.

Another important and widely used natural protein in the textile industry, in the form of fibers, is silk. It can be obtained either from silkworms (the mulberry silkworm *Bombyx* mori, e.g., produces fibers to form its cocoon) or web-weaving spiders such as *Araneus diadematus* or *Nephila clavipes*. The silk achieved from *B. mori* is widely studied and applied in various applications. It is mainly made up of fibroin (diameter = 100 nm; made up of bundles of nanofibrils with a diameter of around 5 nm), with a semicrystalline structure having approximately 55 % β-sheet crystallites. The nanofibril proteins contain three different fractions with different molar masses. The main components are heavy chain (H-chain) fibroin (molar mass is around 350 kDa; Gly-Ala-Gly-Ala-Gly-Ser repeat unit) and light chain (L-chain) fibroin (ca. 25 kDa) with small fractions of a P25 protein (ca. 30 kDa).

Spiders are excellent fiber producers, and their spider nets are ubiquitously present in the environment (Figure 1.4). Spider nets consist of spider silk, with fiber diameters in the range of 1–20 µm, and a core-shell structure has always fascinated humankind with its excellent mechanical properties. Outstanding mechanical toughness as a combination of excellent mechanical strength and ductility makes spider silk a unique material. The elongation at break and tensile strength are 30 % and 1 GPa, respectively. Efforts are undertaken in the area of synthetic fibers, including nanofibers, to mimic such properties.

Figure 1.4: Photograph of a wet spider net.

The most common mineral fibers originate from natural materials, such as asbestos, rock wool, and fiberglass.

All these examples show the beauty of nature in creating a wide spectrum of fibers on a molecular and supermolecular level, in a complex hierarchical way, providing impressive properties. Natural fibers show different cross-sectional morphologies with circular, non-circular and fibrillar structures. Many unique properties of natural fibers can be correlated to their hierarchical supermolecular constitution and inspire scientists to optimize the supermolecular structure and morphologies of synthetic fibers, in order to get appreciable mechanical properties and functionalities.

1.3 Synthetic fibers – polymer fibers produced by technical processes

As pointed out earlier, mankind used fibers that were provided by plants and animals. These fibers were subsequently subjected to further processing. The growing need of course tended to set limits to the availability of such fibers. As the demand for textiles grew, novel areas of applications for fibers arose and as the need for introducing novel functions into fibers such as high stiffness, high strength and specific optical functions became obvious, the interest towards synthetic fibers was directed.

At the end of the 19th century, first efforts were realized to develop materials via synthetic routes from precursors – oil, gas, or coal – to be subsequently shaped to fibers via technical processes. Since then man-made fibers consisted mostly of organic polymers. They have replaced, to a large extent, natural fibers in many applications, including textiles. Man-made fibers offered many advantages, such as ease of processing, reduced costs, and additional functionalities via additives such as stabilizers, nanoparticles, and dyes. Furthermore, property modifications are achieved either by application of more than one material or by modification of the spinning process. The broad range of available materials and corresponding spinning techniques, relying on mechanical stress-induced deformations of melts or solutions, developed over the years, has caused fibers to be – without any doubt – key structural and functional elements in a multitude of technical areas.

It is said that Sir Joseph Swan spun the first synthetic fibers from natural material in the early 1880s, using cellulose liquids. This was followed by the production of other synthetic fibers from nitrocellulose, viscose, and cellulose acetate. In all these examples, the fibers were man made, but the materials used originated from nature and therefore were not synthetic fibers in the true sense.

Today, most of the synthetic petro-based polymeric fibers belong to the polymer class, such as polyamides, polyesters, polyolefins and vinyl polymers like polyacrylonitrile (PAN) (precursor fibers for making carbon fibers). Nylon, chemically belonging to the class of polyamides, is in fact considered to be the first purely synthetic fiber. In the 1930s, Wallace Carothers developed it at DuPont. It served as a replacement for natural fibers such as silk. In early times, the textile applications of nylon as a material for women's stockings, parachutes, and high-strength ropes were shown.

In 1941, highly tough and resilient synthetic polyester fibers [Terylene®, chemical name poly(ethylene terephthalate)] were made by John Rex Whinfield and James Tennant Dickson. Today, a large variety of synthetic fibers, made from polypropylene (PP), polyethylene (PE), PAN, polytetrafluoroethylene (PTFE), aromatic polyamide (aramid), and segmented polyurethanes (elastane fibers) are well established [4, 5]. The chemical structures of common fiber-forming polymers are shown in Chart 1. Some of these fibers are characterized as high-performance fibers because of their very high strength, modulus, thermal stability and chemical resistance, required for applications in composites. Therefore, they are used for example in aerospace industries, civil engineering, constructions, and protective apparels.

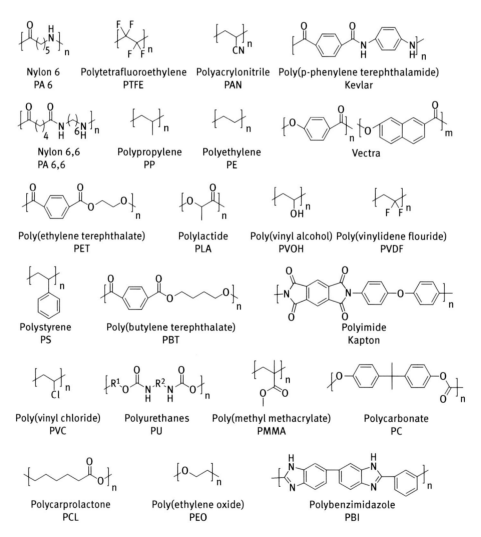

Chart 1: Chemical structure of some representative fiber-forming polymers.

1.4 Spinning of fibers

The first step in the production of synthetic petro-based fibers is the synthesis of polymers from low-molar mass materials called monomers, based on oil, gas, and coal. Depending on the nature of the starting monomers, the polymers can be made by various polymerization techniques, such as condensation polymerization, addition polymerization, vinyl polymerization, or ring-opening polymerization. For details regarding the basics of polymerization techniques, refer to polymer chemistry text books and Chapter 5 of this book [6]. These polymerizations yield long chain molecules (macromolecules), composed predominantly of identical repeat units. Under technical conditions polymers are mostly obtained as pellets. The pellets are converted into a viscous liquid either by melting or dissolving in an appropriate solvent for the processability into fibrous forms. The procedure of forming fibers from liquids or melts is called spinning. The spinning process can be used for making fibers from different organic and inorganic materials, such as polymers and silica-based glasses.

In the first step, the material to be spun has generally to be transformed into a viscoelastic state for example by melting of thermoplastic polymers. Other ways of getting materials into a liquid state is by dissolution or dispersion in an appropriate solvent. For materials that are neither dissolvable nor meltable, a chemical treatment is sometimes helpful in getting soluble or meltable derivatives. In the second step, the viscous solution or melt is passed through small orifices of spinnerets. The term *spinneret* refers to a multipored device through which a viscous melt or solution is extruded to form fine filaments. Molten filaments tend to cool to a rubbery state and subsequently to solidified fibers. However, in the case of solutions the solvent is removed to yield solid fibers. The most important spinning processes being technically used in textile, filter and other industries are based on the process described earlier, with different names. For getting high-strength fibers, a post-drawing step is additionally involved to orient and crystallize macromolecules. The simplest method of fiber spinning from melts is called *melt spinning*. A set of supplementary melt spinning approaches are known as direct spinning and gel spinning. When a melt is directly fed to the spinnerets after a polymer production without palletizing it, the process is called direct spinning. The use of ultra-high molar mass thermoplastic polymers often limits the use of melt spinning because of a very high melt viscosity. In such cases, solution spinning can be helpful. The polymer is dissolved in a suitable solvent to form a viscous solution for passing through a spinneret. The solvent can be removed either by passing it through a nonsolvent bath (wet spinning; generally used for making aromatic polyamide, regenerated cellulose or acrylic fibers) (Figure 1.5) or by evaporating the solvent by utilizing air or inert gas streams (dry spinning) (Figure 1.6). Solvents with high vapor pressures such as tetrahydrofuran, alcohols and acetone are used for in the dry spinning method. Acetate, triacetate, acrylic, or polybenzimidazole fibers are produced by this process.

Gel spinning is a special extension of solution spinning, in which a polymer is dissolved at high temperatures and spun from a gel phase (entangled macromolecular network), generated by cooling the solution. The gel fibers are air-dried and passed through a solvent bath to extract the solvent used for spinning. PE fibers with very high strengths and moduli are made by this process, using special polymer grades [ultra-high-molecular-weight polyethylene (UHMWPE)]. Dyneema® and Spectra® are commercial high-performance polyethylene fibers made by DSM-high-performance fibers, Netherlands, and Honeywell, USA.

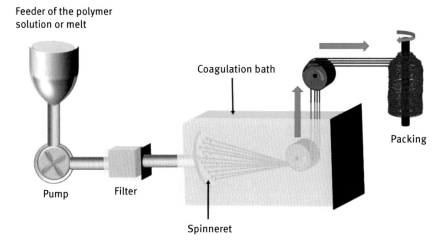

Figure 1.5: Scheme of a setup for wet spinning.

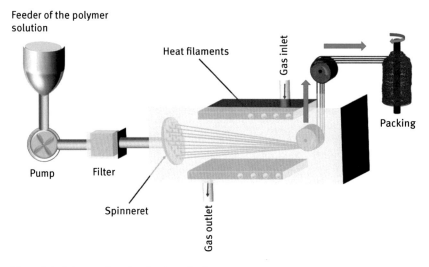

Figure 1.6: Scheme of a setup for dry spinning.

1.5 Fiber structure

The type of spinning process tends to affect the resulting fiber structures and consequently their properties. It leads to orientation of macromolecular chains in the fibers: The chain molecules are stretched along the extrusion (spinning) direction. In the case of glassy polymers, such chain orientations become frozen on solidification and they induce an oriented crystallization with specific crystal planes, becoming preferably oriented along the fiber axis for polymers. So polymers are able to crystallize partially. The orientation causes an increase in stiffness along the fiber axis as well as the onset of birefringence, i.e., the refractive index is no longer isotropic and depends on the direction. The molecular orientation and crystallinity on the surface of fibers is much stronger than that of the interior region. This is caused due to the very high frictional resistance and viscosity during the contact with the wall of a spinneret and radial temperature variations (less on the surface and more in the interior).

Generally, the as-spun polymeric fibers undergo a secondary stretching process, leading to chain orientations along the fiber axis. The draw ratio (the ratio of the initial to the final fiber diameter) is used for the description of the fiber stretching. For example, synthetic fibers such as nylon or high-performance PE fibers are stretched to a draw ratio of 5–7 after the spinning process. The draw ratio significantly influences the degree of macromolecular chain alignments, and hence it alters the crystallinity and mechanical properties in the direction of the chain alignment. They generally increase with increasing draw ratios. Crystallinity can influence the chemical stability, the moisture absorption and the moisture-retaining properties of fibers. The chemical stability increases, but moisture absorption decreases with advancing crystallinity. A higher degree of crystallinity also makes it difficult for a dye to interpenetrate. The fiber diameter is mainly affected by the draw ratio and the size of the spinneret holes. One specific characteristic of most of the fibers coming not only from nature but also from conventional spinning techniques is that their diameters tend to be well within the several micrometers up to tens of micrometer range.

Furthermore, the cross-section of synthetic fibers can be varied by choosing appropriate shapes for the spinneret holes. Circular, dog bone, kidney bean, or trilobal cross-sectional shapes are already reported in the literature. A trilobal cross section of fibers is useful in promoting bulkiness in pile fabrics, such as carpets. The surface texture of fibers can also be varied for synthetic fibers. It depends on the type of material used for spinning, spinning conditions, draw ratio, additives, etc. This topic will also be described later for electrospun nanofibers.

1.6 Fiber applications

In fact, a broad spectrum of fiber applications relate to nonwovens. By the way, these applications also hold for nanofiber-based nonwovens as they will be discussed in Chapter 5. The term *nonwoven* is very often used in the textile industry for fabric sheets containing entangled fibers, which are neither woven nor knitted. Nonwovens display specific properties and functionalities, such as absorbency, resilience, ease of stretching, softness, cushioning, thermal insulation, acoustic insulation and filtration.

Nonwoven materials are used in numerous applications. First of all, those include filter applications for gasoline, oil, air and water, in cases of coffee and tea bags, mineral processing and liquid cartridges, etc. Applications in medical areas include tissue engineering scaffolds, covers and packaging for medical instruments, disposable doctor's gowns, face masks and surgical caps. Mulch membranes, landfill liners and infiltration barriers for drainage tiles and soil stabilizers are further applications of nonwoven structures.

For quite some time, fibers have been utilized for making polymer composites. Both, natural and synthetic high-strength fibers, such as cellulose, jute, glass, and aramid, have been used for the reinforcement of thermoplasts and thermosets. Mechanical properties of fibers, large axial ratios (length/diameter ratio), uniform dispersions and very strong interfaces between reinforcing fibers and polymer matrix are some of the important requisites for making polymer composites. As will become apparent in Chapters 2 and 5, nanofibers offer great potentials in reinforcement applications.

1.7 Interesting to know

- Insoluble and high-melting polymers such as polytetrafluoroethylene (PTFE, common name Teflon) can be spun neither by melt nor by solution spinning techniques. They are spun by a modified method, related to solution spinning, i.e. dispersion spinning. The polymer is dispersed in an aqueous solution, stabilized by tensides like poly(vinyl alcohol) and spun into fibers by wet or dry spinning, followed by thermal decomposition of the tensides at high temperature.
- Liquid crystalline polymers such as poly(p-phenylene terephthalamide) (commonly called Kevlar) and Vectra (aromatic copolyester) (for chemical structures refer to Chart 1) provide high-performance fibers by gel/wet and melt spinning, respectively. This is due to molecular orientation without chain folding during fiber formation due to the rigid mesogens. Kevlar is a lyotrope, showing liquid crystalline behavior in solution (e.g., sulfuric acid), whereas Vectra is a thermotropic polymer which forms anisotropic melts.

– Polyethylene (PE) is available in different thermoplastic grades, such as low-density PE (LDPE), high-density PE (HDPE), linear-low-density PE (LLDPE) and ultra-high-molecular weight PE (UHMWPE). These grades differ mainly in the macromolecular architecture. LDPE is a branched PE, made by high-temperature, high-pressure free-radical polymerization method. HDPE is a linear PE made with the help of metal catalysts. In 1963, Ziegler-Natta won the Nobel Prize for using metal catalysts for ethylene and stereoregular propylene polymerization. LLDPE is a copolymer of ethylene and α-olefins, made by metal-catalyzed vinyl polymerizations with regular side chain branches. The side chain branches come from the side group of α-olefins in LLDPE. UHMWPE has a very high molar mass (more than 2 millions g/mol) because of metallocene-catalyzed polymerization of ethylene.

References

[1] A. K. Mohanty, M. Misra, L. T. Drzal, *Natural fibers, biopolymers, and biocomposites*, CRC Press, Taylor & Francis, Boca Raton, FL, 2005.
[2] R. Kozłowski, *Handbook of natural fibres*, Woodhead Pub., Oxford, Philadelphia, Pa., 2012.
[3] S. Kalia, A. Dufresne, B. M. Cherian, B. S. Kaith, L. Avérous, J. Njuguna, E. Nassiopoulos, *International Journal of Polymer Science*, *2011*, 2011, 1–35.
[4] J. E. McIntyre, *Synthetic fibres. Nylon, polyester, acrylic, polyolefin*, CRC Press; Woodhead Pub., Boca Raton, FL; Cambridge, England, 2005.
[5] W. E. Morton, Hearle, J. W. S, *Physical properties of textile fibres*, CRC Press; Woodhead Pub., Boca Raton, FL; Cambridge, England, 2008.
[6] D. Braun, *Polymer synthesis. Theory and practice: fundamentals, methods, experiments*, Springer, Berlin, New York, 2013.

2 Nanofibers

2.1 Benefits of nanofibers

It is clear that fibrous nonwovens, both natural and synthetic, are not only important for textile applications but also indispensable for many other specialty uses. The application of fibers for a particular request is decided by their material properties, such as mechanical stability, ability to withstand high temperatures, high softening points, chemical resistance, hydrolytic stability, and the costs of production. For many applications such as filter, reinforced composites, tissue engineering, and sensors, the fiber diameter also becomes decisive, which also governs the pore size of nonwovens. For filter applications, the pore size of the nonwoven decides the size of the impurities to be filtered off, as air and wet filtration are mere physical phenomena. Lowering the fiber diameter means reducing the pore size and therefore an increase in the filtration efficiency and filtering out of very small impurities, in both air and liquid filtration. The reduction in the fiber diameter also increases the specific surface area. This leads to improved filter efficiencies in retention of impurities on filter surfaces, or in cases of chemically reactive filters to higher reactivity.

Fiber reinforcement also benefits strongly from very thin fibers with high porosities and large specific surface areas, by making superior wetting and strong interfaces with matrix polymers. Depending on the spinning method, the reduction in fiber diameters could also lead to an abrupt increase in the tensile strength and stiffness, which could be highly beneficial for reinforcement purposes. Figure 2.1 shows a sudden change in the mechanical properties of electrospun polyacrylonitrile fibers when the fiber diameter is reduced below 150 nm due to molecular orientations [1]. Thin fibers, with fiber diameters lower than the wavelength of visible light, would also provide transparent polymer-reinforced composites by avoiding light scattering.

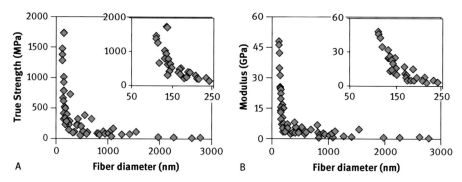

Figure 2.1: Correlation of fiber (polyacrylonitrile) tensile strengths (A) and moduli (B) with the fiber diameter. There is a sudden change in the mechanical properties on decreasing fiber diameters. [Reprinted by permission from American Chemical Society: *ACS Nano* **2013**, *7*, 3324, copyright ACS (2013)] [1].

The application of nanofiber nonwovens as scaffolds for tissue engineering also benefits from low fiber diameters, as they resemble extracellular matrix and the large surface area of nanofibers facilitates binding of proteins and cell membrane receptors during cell culture. Therefore, they could enhance cell proliferation [2].

In addition, decreasing fiber diameters to nanoscales would bring changes in confinement and mass transport properties of nonwovens, which could be of interest for example for applications in drug-release and catalysis.

2.2 Production techniques for thin fibers

The problem is that conventional fiber-processing techniques, briefly described in Chapter 1, are technically not suitable to form nanofibers. In conventional fiber-spinning techniques such as melt and solution spinning, the fiber diameter is mostly controlled by the spinneret hole size, drawing ratio and drawing speed. Submicrometer-sized structures are not really available along these routes, where fiber formation is mainly controlled by mechanical deformation processes, since too many defects would lead to destabilization. Therefore, special techniques have been developed [3] to get very fine, low-defect fibers, which are described in the following section.

2.2.1 Capillary flows

An interesting approach for making very thin fibers is the use of capillary flows, which stretches fluid interfaces (two immiscible fluids, one acts as templating fluid and the other forms the material to be processed to fibers), sucked by mechanical forces through a die, down to micrometer and submicrometer dimensions [4]. The method is suitable for making both particles and fibers in the submicrometer range. Depending on the sucking speed, either one or both layers can be subjected to convergent flow and could lead to core–shell type structures with a template polymer layer outside. The complex fiber-forming processes involved in this technique do not allow larger-scale fiber productions.

2.2.2 Melt blowing

Melt blowing is one of the technical methods for the production of microfibers in large amounts. In melt blowing, a thermoplastic polymer is extruded through small orifices as in conventional melt spinning [5] (Figure 2.2). In the process, the molten strands ejected from the orifices are subjected directly to a convergent stream of hot air. The air stream causes a rapid cleavage of the strands into fibers with smaller and smaller diameters, which are subsequently deposited on a collector. The presence of orifices close to each other gives rise to a complex flow pattern, so that the fibers do not remain parallel to each other and parallel to the machine direction, defined by the orifice and the collector. The overall fiber orientation becomes isotropic at some distance from the orifices and finally perpendicular to the machine direction. Furthermore, the complex flow causes an entanglement of the fibers, which in turn affects the flow pattern and fiber fission. Thus, melt blowing causes the deposition of entangled fibers rather than isolated fibers, which is not necessarily a disadvantage, if the aim is to produce nonwoven architectures. Air speed characteristics of melt blowing are in the order of 100 m/s, causing maximum fiber speeds of around 50 m/s and accelerations of 10^5 m/s². Specific values for fiber attenuation are fiber diameters of about 500 μm close to the orifice, as defined by the diameter of the orifice, and several micrometers down even to 500 nm at the collector. This technique has limitations among others in terms of the processing of fibers with temperature-sensitive additives.

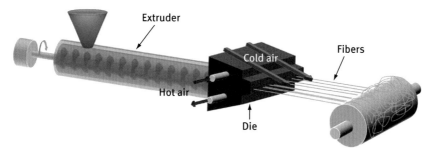

Figure 2.2: Schematic of melt blowing.

Utilization of more than one polymer for spinning can provide specialized morphologies, such as island-in-sea and segmented splittable fibers, which are used for making very thin fibers by removing one component. In island-in-sea-type conjugated spinning, a blend of two polymers is spun, followed by fiber drawing that leads to a special morphology in which many thin fibers from one component are arranged in the second matrix component (Figure 2.3) [6]. The matrix component is considered as sea, and the thin fibers distributed in the matrix are called islands, hence the name island-in-sea is given. The removal of the sea component provides nanofibers of the first component, which was present as islands. Very thin fibers (less than 50 nm) can also be made by this technique.

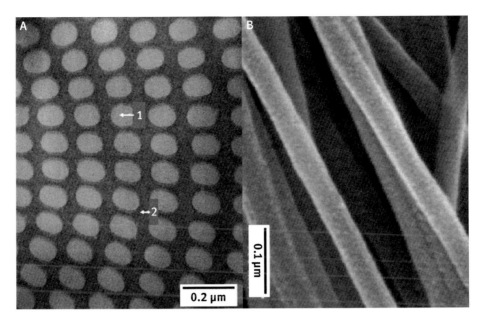

Figure 2.3: (A) Island-in-sea type morphology as observed by TEM in which poly(ethylene terephthalate) (PET) fibers (1) are distributed as islands in nylon-6 matrix (2) as sea; (B) PET nanofibers after removal of matrix component as seen by SEM [6]. [Reprinted by permission from John Wiley and Sons: Poly(ethylene terephthalate) nanofibers made by sea–island-type conjugated melt spinning and laser-heated flow drawing. [Reprinted by permission from *Macromol. Rapid Commun.* **2007**, *28*, 792–795, copyright WILEY-VCH Verlag GmbH & Co. KGaA, Weinheim (2007)].

2.2.3 Solution blowing

Solution blowing is another approach toward submicrometer-sized fibers with high productivity [7]. It makes use of a concentric nozzle. The polymer solution is delivered to the inner nozzle through a pump, and a high velocity gas flows from the outer nozzle. The gas flow helps stretching of the polymer jet, coming out through the inner nozzle by inducing shearing at the interface of gas and polymer solution, by overcoming the surface tension, leading to fine polymer jets directed toward the collector. The solvent evaporates during the flight, and the solid fibers deposit on the collector in the form of randomly oriented fibers, i.e., nonwoven mat (Figure 2.4).

Fiber diameter, morphology and production rate are affected by the polymer solution flow rate, polymer type, gas flow pressure, solution concentration, the distance between the nozzle and the collector and the protrusion distance of the inner nozzle. The solution concentration has a major influence on the fiber diameter, whereas the productivity depends to a large extent on the polymer solution injection rate, the gas flow pressure and the distance between the nozzle and the collector.

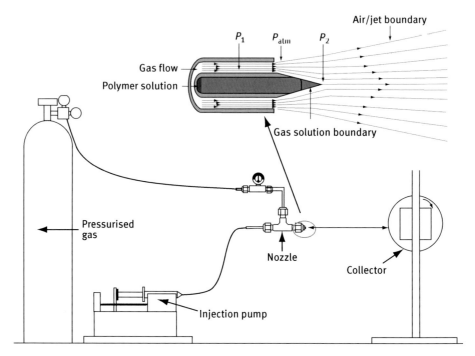

Figure 2.4: Schematic of solution blow spinning and the special nozzle. A polymer solution is pumped through the inner nozzle and high-pressure (P1) gas passes through the outer nozzle. Shearing at the gas – polymer solution interface helps to create low pressures around the inner nozzle (P2) that draws the polymer solution into a jet [7]. [Figure adapted by permission from John Wiley and Sons: Solution blow spinning: A new method to produce micro- and nanofibers from polymer solutions, *J Appl. Polym. Sci.* **2009**, *113*, 2322–2330, copyright © WILEY-VCH Verlag GmbH & Co. KGaA, Weinheim (2009).]

Obviously, technical processes are available, which allow the production of fibers with very fine diameters in the submicrometer range either from the molten state or from solutions. One disadvantage is, however, that these methods do not seem to be able to produce nonwovens composed of long individual fibers, but rather entangled fibers or strands of fibers. Efforts to orient such fibers in a highly regular way were not accomplished. The controlled production of fibers with the following characteristics was not readily possible: close to infinite lengths; beaded, undulating diameters; non-circular cross section; specific surface features such as branches or barbs; functionalities induced by the inclusion of biological objects such as cells, bacteria, proteins and growth factors; and diameters really down to a few nanometers. In contrast, electrospinning can fulfill these demands. It is a roll-to-roll process for the production of nanoscaled materials which makes it unique in combination with the above mentioned demands. Additionally, electrospinning allows near endless materials combinations which is otherwise restricted with alternative nanofiber production processes mentioned before. Consequently, electrospinning is the state-of-the-art techique for the preparation of nanofibers as described in the next section.

2.3 Nanofibers via electrospinning

Electrospinning utilizes a strong electric field for overcoming the surface tension to produce a spinning jet from solutions, suspensions or melts [8, 9]. This technique manufactures continuously long nanofibers, nanowires and nanotubes. The process provides easy access to fibers with specific surface topologies to specific assemblies of nanofibers, such as randomly oriented fibers (nonwovens) or orientationally aligned fibrous nonwovens, depending on the specific electrode configurations used as collectors. Electrospinning provides not only polymer-based (natural and synthetic polymer) fibers but also gives access to inorganic, metal and ceramic nanofibers. Furthermore, electrospinning allows the functionalization of the nanofibers during preparation, by incorporating features such as viruses, bacteria, enzymes, catalysts, drugs, metal nanoparticles, nanowires, and nanotubes.

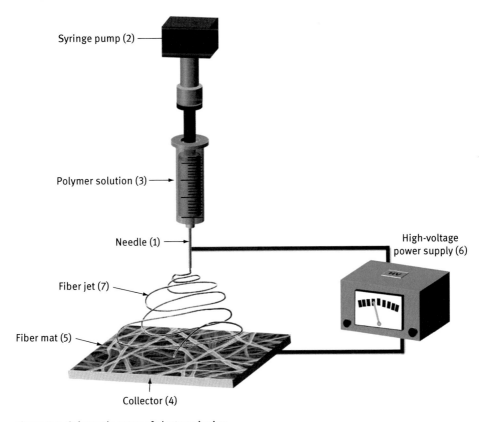

Figure 2.5: Schematic setup of electrospinning.

2.3.1 Fibers by electrospinning using a needle-type electrospinning setup

The schematic of electrospinning is shown in Figure 2.5.

The polymers used for electrospinning should be either soluble in a solvent or meltable. The solvent for dissolving the polymers can be either water or any organic solvent. The simple electrospinning machine has the following components:

Syringe with a metallic needle **(1)**: The plastic doctor's syringe of different capacities (1 ml, 2 ml, etc.) can be used as reservoir for the spinning solution. Instead of syringes, special nozzles can also be applied. Metallic syringes with surrounding heating jackets are generally utilized for melt-electrospinning. A metallic needle with an inner diameter of around 100 μm attached to the syringe acts as both: a die and the primary electrode, to which high electric fields in the range of typically 100–500 kV/m are applied. The productivity can be enhanced by the use of multiple needle systems. However, the use of multiple needle systems could lead to inhomogeneity in the nonwovens during fiber deposition which can be overcome by movement of the needles (1) or of the collector (4).

Feeding pump **(2)**: The spinning liquid **(3)** (melt/solution) is pumped through the needle **(1)** with a particular feeding rate, using a piston/peristaltic pump. The feeding rate can vary between a few microliters per hour up to several milliliters per hour depending on the viscosity of the melt/solution and the type of reservoir used.

Collector **(4)** and secondary (counter) electrode: The collector can either act as secondary electrode (in electric contact) or is arranged in the neighborhood of the counter electrode (with a different potential). Generally, it is placed 10–25 cm away from the primary electrode, but this distance can also be more or less, depending on the spinning situation. The collectors can be stationary or rotating. The randomly oriented fibers **(5)** are collected on stationary collectors, whereas rotating collectors can provide aligned fibers (depending on the rotational speed). Although aluminum foil or glass plates are the most commonly used collectors, metallic frames, drums, filters, disks and backing paper can also be used at controlled temperatures. Even liquids can be used as collector.

High-voltage supply **(6)**: For making fibers, a high voltage (generally 1kV/cm) is required between the primary and the counter electrode. The electrical current should only be a few hundred nanoamperes up to several microamperes.

The fiber formation **(7)** by electrospinning takes place in different steps. To start the spinning process, the polymer solution or melt is filled in the spinning nozzle/syringe equipped with a metallic needle and a high electric field is applied between the tip of the nozzle and the collector or counter electrode. The polymer droplet formed at the tip of the needle will be electrostatically charged, and the conical deformation of the droplet (Taylor cone) leads to the onset of jetting. Commonly, a single polymer jet is formed, followed by the development of a rectilinear jet and bending deformations with looping, spiraling trajectories, etc. In addition, processes exist that also affect fiber geometries, such as Rayleigh instabilities, the electrically driven axisymmetric instabilities, causing diameter undulations as well as branching processes.

Electrostatic forces and the surface tension play an important role in the jetting process, whereas gravitational forces are not important for making fibers by electrospinning. A stable jet is formed when electrostatic repulsive forces between charged elements of the liquid jet overcome the surface tension. The number of jets formed per droplet is dependent upon the balance between the voltage applied and the surface tension. So sometimes more than one electrospinning jet can be formed from a single droplet. The charged jet accelerates toward the counter electrode and thins due to elongation and evaporation of the solvent. This leads to the deposition of fibers on the counter electrodes or collectors. The jet coming out from the tip of the droplets in the beginning follows a straight path for a very short distance followed by a bending, winding, curling and looping. The loop diameter increases whereas the jet diameter decreases along its path. The reduction in diameter is due to the evaporation of the solvent and the longitudinal deformations of the jet induced by electric forces. Strain rates of the order of 1,000 s^{-1} and elongational deformations of up to 1,000 are characteristics of electrospinning as far as the linear part of the jet is concerned, while the bending instability can increase these values by up to a factor of 100, leading to strong molecular chain extensions [8a]. The fibers made by electrospinning are generally infinitely long, with diameters ranging from several micrometers down to a few nanometers. The orientation of fibers is controlled by the type of collector used. Randomly oriented fibers are collected on a stationary collector. They are also called nonwovens or nanomats. Rotating disks, drums and metallic frames provide aligned fibers. Some of the common collectors used in electrospinning are shown in Figure 2.6, and randomly oriented and aligned fiber formation is shown in Figure 2.7.

For the characterization and observation of the spinning process an optical/electron microscope (Chapter 3.2.1) and a lamp are needed. The light helps by observing the fiber formation during the spinning process, whereas the microscope gives information about the size and morphology of the resulting electrospun structures. Optionally, the spinning process can be observed using high speed camera. When it comes to the modulation of the fitting value for the electric field, the applied voltage is the easiest way to modify the field. For a first trial, the adjustment of the voltage can be done by applying 1 kV onto the collector and slowly increasing the voltage of the needle. There is a certain range of voltage where the fiber formation is stable. Before that point, the droplet at the tip expands (*Taylor cone*), whereupon discontinuous fibers are forming (start of fiber splitting), which is often accompanied by small droplets. A too high voltage finds expression in fiber breakages or even again in droplet formation, but those droplets are discernibly smaller and can often be recognized only by using a microscope (electrospraying; Figure 2.8). Another way to define the electric field is to vary the distance between the needle and the collector, but this value is also associated with smaller times for the solvent to evaporate (fibers could stick together). This is again accompanied by the pathway of stretching of the fibers, since the alignment of the macromolecules and the variation of the diameter of the fibers can proceed only if the molecules are able to move. However, the last point counts for every increase of the electric field.

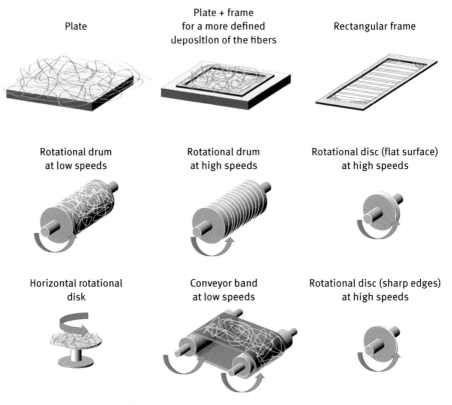

Plate

Plate + frame
for a more defined
deposition of the fibers

Rectangular frame

Rotational drum
at low speeds

Rotational drum
at high speeds

Rotational disc (flat surface)
at high speeds

Horizontal rotational
disk

Conveyor band
at low speeds

Rotational disc (sharp edges)
at high speeds

Figure 2.6: Schematic of different collector types for electrospinning. Rotating drums and disks provide aligned fibers, and the speed of rotation controls the degree of fiber alignment. Aligned fibers can also be prepared with rectangular frames, without any rotations of the collector. Slow rotating drums and horizontal disks or plates with an additional frame are often applied for the production of highly uniform fiber mats with randomly oriented fibers.

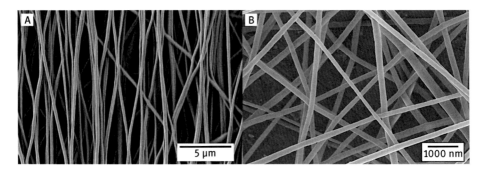

Figure 2.7: SEM pictures of aligned and randomly oriented electrospun poly(vinyl alcohol) (PVOH) fibers. Flat stationary collector provides randomly oriented fibers (*right*) as compared to a rotating disk that gives aligned fibers (*left*). The degree of orientation depends on the rotating speed.

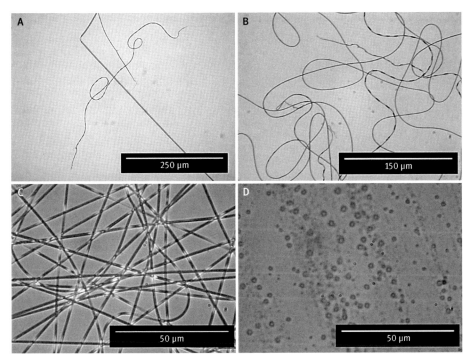

Figure 2.8: Optical microscope images of PVOH fibers, electrospun with different voltages from a 6 wt% solution (87–89% hydrolyzed PVOH; Mw (weight average molar mass) = 146,000–186,000 g mol^{-1}). (A) 355 V cm^{-1}; (B) 555 V cm^{-1}; (C) 855 V cm^{-1}; (D) 1,500 V cm^{-1}. For the calculation of the electric field, a parallel-plate capacitor is assumed with a diameter of 1 cm.

In the following table, some hints are listed for what can be done in typical situations facing problems during electrospinning.

Problem	Increase the following values	Decrease the following values
Formation of beads (beaded chains)	Electric field (e.g., voltage of the needle); molecular weight of the polymer; weight concentration of the polymer; conductivity of the spinning solution (addition of conducting salt); temperature	Surface tension of the solution (e.g., addition of surfactants); humidity
Deposition of fragments (spinning solution seems more to be extruded out of the syringe)	Electric field (e.g., voltage of the needle); conductivity of the spinning solution (addition of conducting salt)	Weight concentration of the polymer; flow rate of the spinning solution

(continued)

(continued)

Problem	Increase the following values	Decrease the following values
Deposition of fragments (fragments are inside of droplets)	Homogeneity of the spinning solution (longer stirring times or use other mixing systems as vortex mixers, KPG stirrer or elevated temperatures during the stirring)	
Deposition of fragments (fragments are built at the tip of the needle)	Electric field (e.g., voltage of the needle); flow rate of the spinning solution	Weight concentration of the polymer (or use/add other solvents with lower vapor pressure); temperature
Spraying/no fibers at all	Molecular weight of the polymer; weight concentration of the polymer	Electric field (e.g., voltage of the needle)
Fibers are climbing up from the collector to the needle	Conductivity of the spinning solution (add the conducting salt, change the solvent, or use a rotational collector)	Weight concentration of the polymer
Discontinuous fiber formation	Electric field (e.g., voltage of the needle); flow rate of the spinning solution; homogeneity of the spinning solution (longer stirring times or use other mixing systems as vortex mixers, KPG stirrer or elevated temperatures during the stirring)	Surface tension of the spinning solution (e.g., addition of surfactants)
Forming of flat ribbons	Distance between the needle and the collector (or add/use solvents with higher vapor pressures)	Molecular weight

Although the most frequent arrangement of electrode and counter electrode is top-down, providing spinning from top to bottom, other arrangements such as bottom-up and horizontal (sideways) are also possible, since the driving force of the fiber formation is charge balancing and not the gravitational forces. The advantage of spinning from bottom-to-up or sideways is the elimination of polymer drops falling on nonwovens during the spinning process, which are often seen due to the instability of the electrospinning process.

Single-jet electrospinning is a very slow process with throughputs in the range from 10 µl/min to 10 ml/min. Assuming a feeding rate of 1 ml/h for a polymer solution, characterized by a polymer concentration of 20 wt%, the amount of fibers deposited in 1 h on the deposition plate corresponds to 0.2 g or to a nonwoven volume of slightly below 1 cm³, as controlled by the degree of porosity of the nonwoven (mostly around 0.8) and of the density of the polymer (often around 1.1–1.2 g/cm³). Therefore, similar setups with multiple syringes can be employed either to increase the productivity or to introduce additional functionalities by using different polymers and/ or additives.

Many other modifications are available in the literature offering much larger production capacities. The special needleless electrospinning arrangements, such as porous walled cylindrical tubes filled with a polymer solution form many jets at the same time increasing productivity. Here, multiple drops are generated on the outer surface of the tube, when the polymer is pushed through the pores. Each drop act as jetting points when the solution is charged [10]. In 2009, Wang et al. [11] demonstrated a conical metal wire-coil for the spinning of polymer solutions. Multiple jets are formed on the coil surface without using any solution channels. Lu et al. [12] introduced a very high-throughput needleless electrospinning by using a rotary cone. A production that was several thousand times higher than that by single-needle electrospinning was reported. Another interesting approach was the combined use of magnetic and electric fields acting on a two-layer system [the two layers being the polymer solution (upper layer) and a ferromagnetic suspension (lower layer)]. The perturbations were generated at the polymer surface after the application of a normal magnetic field and provided multiple electrified jets [13]. A 12-fold enhancement of the production rate was reported in contrast to the conventional single-needle electrospinning. Wu et al. [14] reported a high throughput of electrospun nanofibers by using a circular cylinder for generating multiple fiber jets. A production of nanofibers of more than 260 times in weight compared to that of conventional single spinneret electrospinning was demonstrated. Spinning from agitated surfaces and roller-type electrodes, from coil-shaped electrodes and from bubbles emerging from the spinning solution was also tested. It should be sufficient here to point out that the basic fiber-spinning processes involved in all these methods remain the same. The details of different types of fiber generators in needleless electrospinning (Figures 2.9 and 2.10) can be referred in Ref. 15.

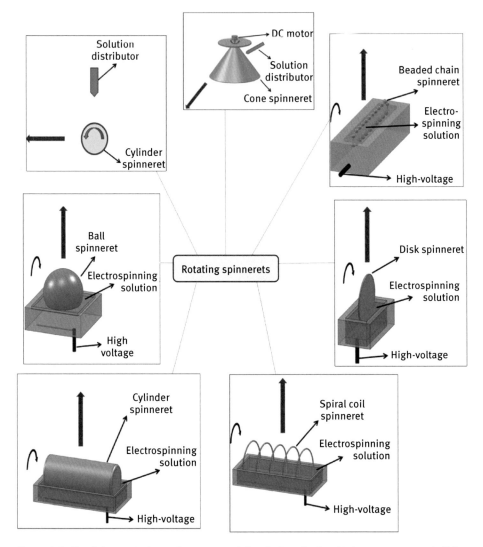

Figure 2.9: Needleless, rotating spinnerets used for electrospinning. Red arrows are symbolizing the spinning directions [15]. (Figure is taken from an open access article published by H. Niu, T. Lin, *J. Nanomater.* **2012**, *2012*, article ID 725950.)

The result of all these developments was the change of perspective on electrospinning from being a coating technique to a membrane-forming process. Although electrospinning can be used for coating any liquid/solid substrate under batch conditions or by roll-to-roll process, it can now-a-days produce many meters of continuous self-supporting membranes in different widths. Even yarn-type fibers are possible by electrospinning using special electrodes (Figure 2.11).

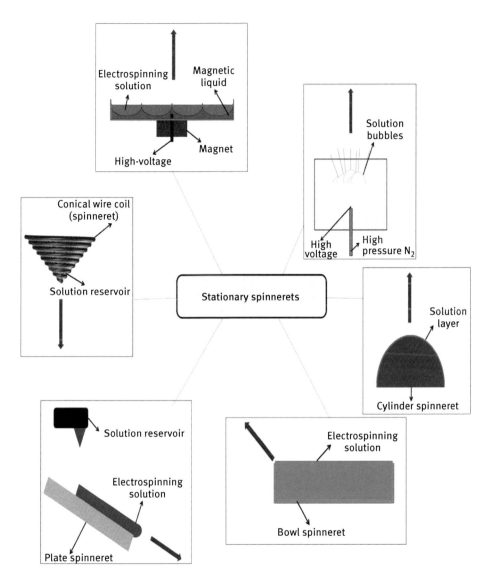

Figure 2.10: Needleless, stationary spinnerets used for electrospinning. Red arrows are symbolizing the spinning directions [15]. [Figure is taken from an open access article published by H. Niu, T. Lin *J. Nanomater.* **2012**, *2012*, article ID 725950.]

2.3.2 Bicomponent fibers by electrospinning

Furthermore, to get special morphologies and encapsulations of additives, drugs, particles, bacteria, virus, etc., special spinning nozzles were developed, such as side-by-side, coaxial or triaxial types. As the name indicates, two or more spinning solutions are allowed to flow in special nozzles with either two chambers arranged side-by-side, separated with a membrane (side-by-side nozzle) or by core–shell geometry (coaxial nozzle). After the first use of coaxial electrospinning by Greiner and Yarin et al. [16] for the formation of poly(dodecylthiophene) and poly(ethylene oxide) core–shell fibers (Figure 2.12A), the technique has gained tremendous attention not only for the encapsulation purposes but also for making hollow nanofibers for use as nanoreactors, as carrier for catalysts, etc. Interestingly, the nonspinnable polymers (might be attributable to the low molar mass) can also make core fibers, due to the template effect from the shell spinnable material. Even non-fiber-forming metal salts can be arranged in the form of 1D wires using core–shell spinning. The example was shown by Greiner and Yarin et al. [16] by taking palladium acetate [Pd(OAC)$_2$] as core solution and poly(L-lactide) as shell material. An annealing of the electrospun fibers at 170 °C for 2 h provided Pd 1D nanowires with a diameter of around 60 nm as core (Figure 2.12B).

Figure 2.11: The technique of electrospinning is used (A) for coating purposes and in making (B) self-supporting porous membranes and (C) fiber yarns.

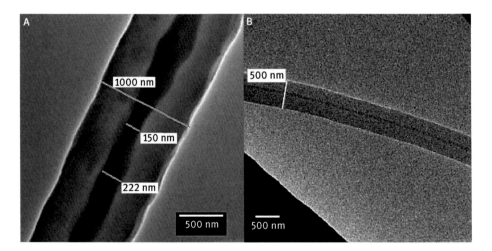

Figure 2.12: TEM images of core–shell fibers. (A) Poly(dodecylthiophene) as core and poly(ethylene oxide) as shell; (B) palladium metal as core and poly(L-lactide) as shell [16]. [Reprinted with permission from *Adv. Mater.* **2003**, *15*, 1929–1932. Copyright WILEY-VCH Verlag GmbH & Co. KGaA, Weinheim (2009).]

A further variation of coaxial spinning is triaxial spinning, with a core–shell–shell morphology (Figure 2.13) [17, 18]. The technique offers advantages when a material is sandwiched between two layers for self-assembly, reinforcement purposes, and encapsulations of more than one material with different release profiles.

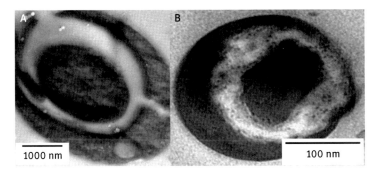

1000 nm 100 nm

Figure 2.13: Triaxial morphology as seen by TEM for (A) polystyrene – thermoplastic polyurethane – polystyrene [Reprinted with permission from *ACS Appl. Mater. Interfaces* **2014**, *6*, 5918–5923. Copyright (2014) American Chemical Society] [17] and (B) polystyrene-block-polyisoprene copolymer-magnetite nanoparticle sandwiched between silica layers. [Reprinted with permission from Small **2009**, *5*, 2323–2332. Copyright WILEY-VCH Verlag GmbH & Co. KGaA, Weinheim (2009)] [18].

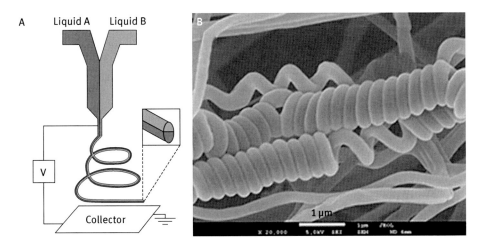

Figure 2.14: (A) Schematic of electrospinning setup with special side-by-side nozzle for making Janus fibers [Reprinted with permission from *Nano Lett.* **2007**, *7*, 1081–1085. Copyright (2007) ACS] and (B) nanosprings produced by side-by-side electrospinning using Nomex and thermoplastic polyurethane [Reprinted with permission from *Macromol. Mater. Eng.* **2009**, *294*, 265–271. Copyright WILEY-VCH Verlag GmbH & Co. KGaA, Weinheim (2009)] [20].

Electrospinning offers excellent opportunities to combine polymer properties in a single system by arranging two materials in a parallel side-by-side morphology, in a simple way [19]. Such fibers are also named Janus fibers. This appearance is achieved by the simultaneous spinning of two polymer solutions, using special side-by-side nozzles (Figure 2.14A). The right choice of materials for the two sides of the fibers and spinning conditions can also provide further interesting morphologies such as nanosprings (Figure 2.14B) and Janus particles [20].

It is of general importance to point out here the major difference between fiber formation by electrospinning and other conventional techniques like melt spinning in which mechanical forces and geometric confinements such as imposed by dies control fiber spinning. Fiber formation in electrospinning is rather controlled by repulsive Coulomb interactions between charged elements of the fluid body, inducing self-assembly processes which follow the general Earnshaw theorem of electrostatic. According to this theorem it is impossible to prepare stable fluid structures such as stable fluid jets in which all elements interact only by Coulomb forces. Charges located within the fluid jet, in the case considered here, move the polymer elements to which they are attached along complex pathways in such a way that the Coulomb interaction energy is minimized as far as possible. Droplet deformation, jet initiation, and in particular, the bending instabilities are predominantly controlled by this kind of self-assembly principle. Highly complex fiber architectures, such as fibers with vertical protrusions, and splayed fibers are further manifestation of it. So while in the case of supramolecular structures, attractive forces such as hydrogen bonding, charge transfer interactions etc. direct self-assembly processes, in electrospinning this role is assumed by repulsive forces. In the last 10–15 years, a lot of progress has been made in theoretical and experimental understanding of the electrospinning process, with an exponential increase in the number of publications and patents. Some of the recommended reviews and books for further reading regarding applications of electrospun fibers are given as references [21–25].

2.3.3 3D and patterned fibrous structures

Conventional electrospinning provides 2D fibrous nonwovens with continuously long fibers. Modified electrospinning methods using special collectors and nozzles can provide 3D and patterned fibrous structures. The use of salt particles and cooled collectors might be helpful in getting 3D structures. The simplest way of getting 3D nonwovens would be to make layer-by-layer electrospinning. For spinning solutions with high conductivity the repulsion of charges in electrified jets would provide 3D fluffy fibers. The different layers can be of the same polymer or different materials. The electrospinning of polyacrylonitrile mixed with multiwalled carbon nanotubes provided such fluffy nanofibers due to poor discharge and interlayer electrostatic repulsion. The use of an additional metal ring auxiliary electrode led to the collection of fibers under the nozzle in a 3D fluffy layer-by-layer form (Figure 2.15) [26].

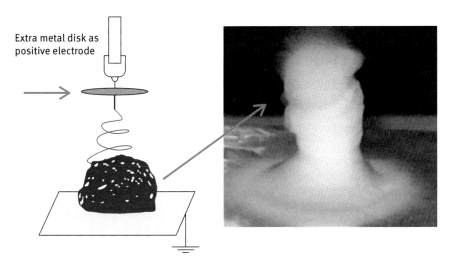

Figure 2.15: Collection of electrospun fibers in a 3D fluffy form due to high conductivity of spinning solution and use of an auxiliary electrode for concentrating the fibers under the nozzle. [Reprinted with permission from *J. Eng. Fibers Fabrics* **2012**, *7*, 17–23. Copyright *J. Eng. Fibers Fabrics*, P.O. Box 1288, Cary, North Carolina 27512-1288, USA (2012).]

The surface resistivity of materials can also be reduced for getting 3D electrospun mats by adding appropriate amounts of surfactants. For example, electrospinning of 25 wt% zein from 70% v/v aqueous ethanol provides 2D nanomats, whereas an addition of 25 wt% SDS (sodium dodecyl sulfate) to the spinning solution can provide 3D mats with very high porosities (Figure 2.16) [27].

Figure 2.16: Three-dimensional (*left*) and two-dimensional (*right*) zein electrospun fiber mats. [Reprinted with permission from *Langmuir* **2013**, *29*, 2311–2318. Copyright American Chemical Society (2013).]

Such fluffy structures are not very useful for applications where 3D is of utmost importance, such as scaffolds for tissue engineering and electrodes for microbial fuel cells, as they do not possess any compressive strength and can be easily compressed to 2D structure.

Another way would be to simply sinter many 2D electrospun layers together to fabricate 3D structures. The sintering temperature is limited for polymeric electrospun fiber mats as they will lose their form at temperatures higher than the glass transition and melting temperatures. Moreover, the sintering temperature is also limited by low degradation temperatures of the polymers; most of the polymers start degrading in the range of 300–400 °C. In such cases, pressurized gas can be applied for sintering fibers, which are in contact with each other, by lowering the glass transition temperature (in this case the gas operates as plasticizer). The stacked and sintered layers of electrospun poly(lactide-co-glycolide) (PLGA) and hydroxyapatite (HA), using pressurized gas (400 psi) are shown in Figure 2.17 [28].

Figure 2.17: Stacked and sintered layers of poly(lactide-co-glycolide) (PLGA) and hydroxyapatite (HA) with 0, 5, 10 and 20% of HA (from left to right). [Reprinted with permission from *J. Polym. Sci.: Part B: Polym. Phys.* **2012**, *50*, 242–249. Copyright WILEY-VCH Verlag GmbH & Co. KGaA, Weinheim (2011).]

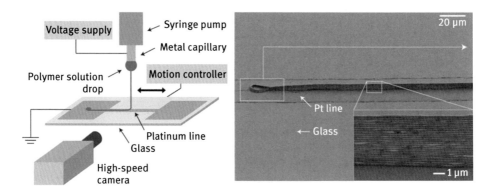

Figure 2.18: Schematic of making 3D layer-by-layer pattern using 20 μm platinum line on a glass plate for focusing the electrified jet (*left*) and SEM of the corresponding 3D structure (*right*). [Reprinted with permission from *Langmuir* **2014**, *30*, 1210–1214. Copyright American Chemical Society (2014).]

Layer-by-layer piling of electrospun nanofibers leading to the formation of 3D pattern can also be done by focusing the electrified jet onto the microline (20 µm platinum line on the glass plate) of a metal electrode (Figure 2.18) [29].

An indirect approach for making 3D highly porous (porosity more than 99%, with densitities as low as 3 mg/cm³) structures of different shapes and sizes with very good mechanical compressibility and bending is by utilization of a dispersion of short electrospun fibers (aspect ratio 120–150) and freeze-drying (Figure 2.19). Electrospinning provides continuously long fibers. They can be cut in length by mechanical means in large amounts, dispersed in appropriate solvents, and freeze-dried to get 3D sponge-type structure in any shape and size. The method is easily upscalable [30].

Figure 2.19: Three-dimensional porous structures with very good compressibility and bending ability, made using a dispersion of short electrospun fibers and freeze-drying. [Reprinted with permission from *Adv. Funct. Mater.* **2015**, *25*, 2850. Copyright WILEY-VCH Verlag GmbH & Co. KGaA, Weinheim (2015).]

Three-dimensional hollow tubes of different diameters can be directly fabricated by electrospinning on rotating wires/rods. After the fiber deposition for a definite period of time the electrospun tubes are carefully pulled out from the collector. The time of fiber deposition decides the wall thickness of the tube, whereas the diameter of the wire/rod collector will influence the internal diameter of the tube. The method is not suitable for making tubes with very small diameters (<0.3 mm) due to handling problems. The application of static 3D collecting templates with combinatorial electric fields is one of the easy ways for making tubular structures of different diameters and lengths. Even interconnected tubes can be made [31]. The application of patterned collectors provides the corresponding patterned tubes (Figure 2.20).

Nanoimprint lithography is also used for making patterned fibers. The fibers are textured with suitable nanolithographic gratings, which have been utilized for tailoring the polarization properties of the light-emitting conjugated polymer [poly(9,9-dioctylfluorenyl-2,7-diyl)-co-(1,4-benzo-{2,1,-3}-thiadiazole)] [32].

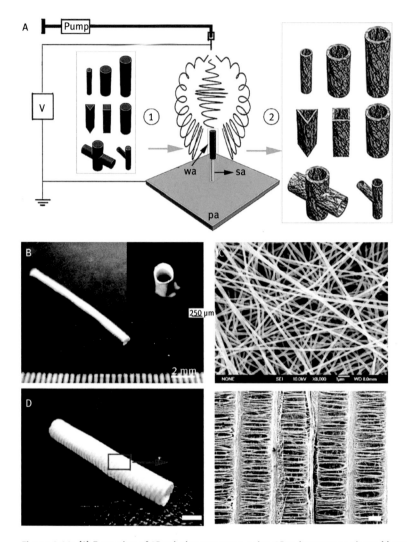

Figure 2.20: (A) Formation of 3D tubular structures using 3D columnar templates (1) and combinatorial electric fields. The 3D fibrous tubes (2) get the shape of the columnar templates. The fibrous tube with 500 μm diameter (B) with random orientation of fibers as seen in SEM (C). The patterned collectors allow the formation of patterned tubes (D, E). [Reprinted with permission from *Nano Lett.* **2008**, *8*, 3283–3287. Copyright American Chemical Society (2008).]

2.4 Interesting to know

– Although the technique of electrospinning dates back at least to the year 1900, an extraordinary attention is provided only after the early 1990s with the work of Prof. Reneker [33, 34]. After a slow start, a dramatic increase in the publications and patents in the field of electrospinning is obvious, dealing with not only the spinning process itself but also with the characterization of the resulting fibers, nonwovens and applications in technical areas and in medicine, with tissue engineering and filter applications being prominent examples. The very first commercial application of electrospun fibers was in filter materials. The word *electrospinning* as entered in ScienceFinder provides 6 hits between the years 1990 and 1995, which increased to 33 in the years 1996–2000 followed by 1,217 for the time period 2001–2005. Amazingly, this number has increased to 21,805 in the last ten years (2006–2014). It is rightly said that electrospinning is like "reinventing the wheel," as many electrospinning setups currently used, including multiple-nozzle spinning and parallel electrodes, are based on design concepts of old patents. The first electrospinning patent was filed in 1900 by John Francis Cooley. In the time period from 1934 to 1944, a number of patents were filed by Formhals, regarding nanofiber yarns based on electrospun fibers, using special nonspinneret electrodes. Worth mentioning is also the early work of N. D. Rozenblum and I. V. Petryanov-Sokolov, generating electrospun fibers in 1938 for filter applications, and Sir Geoffrey Ingram Taylor's work on the mathematical modeling of the shape of the cone, formed by the fluid droplet under the influence of an electric field. That is how it got its name Taylor cone. The review article titled "The history of the science and technology of electrospinning from 1600 to 1995" written by Tucker et al. [35] is a very good document for looking back into the historical developments.
– The methods described till now for the formation of thin fibers are top-down approaches, which are technically used methods. Nanofibers can also be prepared by bottom-up methods, such as self-assembly of small molecules at the molecular level. Figure 2.21A and B shows a self-supporting macroscopic supramolecular nanofiber mat, made by self-assembly of amphiphilic 4-*N*-octanoyl-aminobenzoic acid sodium salt [36]. Very thin nanofibers (<10 nm diameters) can also be achieved by a self-assembly of peptide amphiphiles with β-sheets, oriented parallel to the long axis of nanofibers [37].

Thermally induced phase separation is another bottom-up approach to make nanofibers that is based on the thermodynamic demixing of a homogeneous polymer solution (Figure 2.21C). The demixing is normally achieved by either addition of an immiscible solvent or by cooling the solution below a binodal solubility point [38]. Cellulose nanofibrils are also produced by bacteria *Acetobacter xylinum* in a complex bottom-up process [39]. The bottom-up approaches have limitations in terms of material choice, expensive procedures and scaling up.

Figure 2.21: Bottom-up approaches for making nanofibers by self-assembly of (A) 4-N-octanoyl-aminobenzoic acid sodium salt [Reprinted with permission from *Soft Matter* **2011**, *7*, 1058–1065. Copyright Royal Society of Chemistry (2010)] [36] (B) peptides [Reprinted with permission from *Soft Matter* **2007**, *3*, 454–462. Copyright Royal Society of Chemistry (2007)] [37] (C) thermally induced phase separation of poly(L-lactide) (PLLA) at a gelation temperature of 8 °C. [Reprinted with permission from *J. Biomed. Mater. Res.* **1999**, *46*, 60–72. Copyright (1999) John Wiley & Sons, Inc.] [40].

– Nanofibers of limited lengths can also be made by templating techniques using pores of membranes with definite dimensions as sacrificial templates. The most commonly used templates are porous aluminum oxide membranes. They can easily be made by electrochemical oxidation of aluminum with nanopores of varied diameters (25–400 nm) and pore depths (few 100 nanometers to several 100 micrometers). The polymeric melt penetrates the pores, and an etching of the template provides nanofibers. Mechanical drawing of fibers from polymer melts, vapor phases and chemical oxidative polymerizations are other methods for making nanofibers (Figure 2.22).

– Electrospinning combined with chemical vapor deposition technique can be used for making hollow nano- and mesofibers, as shown for the first time in open literature in the year 2000 [41]. The process involves spinning of a sacrificial/template polymer, which is easily pyrolysable at low temperatures, followed by coatings with a thermally stable polymer, providing a core–shell structure. The removal of the core by heating at a temperature above the degradation temperature of the core polymer provides hollow fibers (Figure 2.23). The process is named as TUFT process (tubes by fiber templates).

– The similar concept can be applied for making hollow fibers by choosing core and shell materials differing in solubility. For example, if you choose a core polymer, which is soluble in water and a shell polymer, which is stable in water, then the core can also be removed simply by dissolution in water. Poly(p-xylylene) (PPX) is a highly thermally stable polymer, which can be used as outer shell, made by chemical vapor deposition. The chemistry of PPX is described in Chapter 4. Using a similar concept, hollow fibers can also be made by dissolution/pyrolysis of the core material in a core–shell fiber made by coaxial spinning.

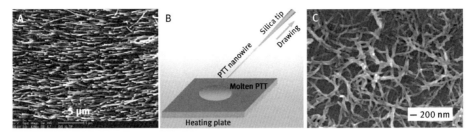

Figure 2.22: (A) Polycaprolactone nanofibers made by a templating technique. Porous aluminum oxide was used as sacrificial template membrane. [Reprinted with permission from *Nano Lett.* **2007**, *7*, 1463–1468. Copyright American Chemical Society (2007).] (B) Schematic for making nanofibers by mechanical drawing of molten poly(trimethylene terephthalate) (PTT) [Reprinted with permission from *Nano Lett.* **2008**, *8*, 2839. Copyright American Chemical Society (2008).] (C) Polyaniline nanofibers made by chemical oxidative polymerization [Reprinted with permission from *Accounts Chem. Res.* **2009**, *42*, 135–145. Copyright American Chemical Society (2009).]

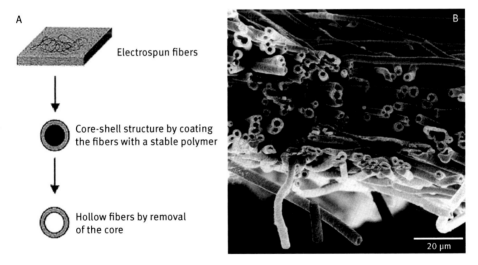

Figure 2.23: (A) Schematic for the formation of hollow fibers by TUFT process and (B) SEM image of hollow poly(p-xylylene) (PPX) tubes. Electrospun poly(L-lactide) fibers were used as sacrificial template fibers, which were removed by heating at 280 °C for 8 h. [Reprinted with permission from *Adv. Mater.* **2000**, *12*, 637–640. Copyright WILEY-VCH Verlag GmbH & Co. KGaA, Weinheim (2000)] [41].

The same method can also be used for making metal hollow tubes, by coating core fibers with a metal via physical vapor deposition (PVD), followed by the coating of the shell material with a CVD. The removal of the core fibers (conducted either by pyrolysis or by dissolution in an appropriate solvent) will provide metal hollow tubes coated with a polymer shell.

— Formation of inorganic, side-by-side Janus fibers (Figure 2.24) is also possible by electrospinning, which was shown by Liu and Sun et al [42].

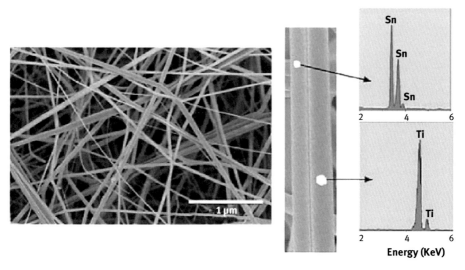

Figure 2.24: SEM image of inorganic bicomponent (TiO_2/SnO_2) nanofibers, as proved by EDS analysis. [Reprinted with permission from *Nano Lett.* **2007**, *7*, 1081–1085. Copyright American Chemical Society (2007)] [42].

– The first use of a triaxial nozzle (Figure 2.25) was shown by Lallave et al. in 2007 [43]. Interestingly, they did not make fibers with core–shell–shell morphology, but used the triaxial configuration of the spinneret to get rid of the problem of precipitations/ solidifications of the polymer at the tip of the nozzle, by fast evaporations of low-boiling solvents. To avoid the solidification, the authors let a solvent flow in the outer sheath, in this case, ethanol. This inhibited the solidification of the Taylor cone.

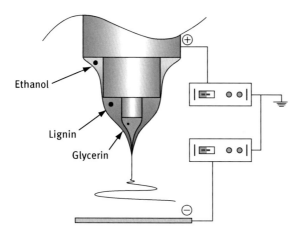

Figure 2.25: Schematic of a triaxial configuration, used for avoiding solidifications of the Taylor cone while electrospinning, by application of low-vapor-pressure solvents. [Reprinted with permission from *Adv. Mater.* **2007**, *19*, 4292–4296. Copyright WILEY-VCH Verlag GmbH & Co. KGaA, Weinheim (2007)] [43].

‒   Generally, the electrospun fibers are smooth without pores. The additional porosity on the surface of fibers can be generated in different ways. The use of low-boiling solvents such as methylene chloride and high humidity during spinning can provide porous fibers. The fast evaporation of the solvent lowers down the temperature on the surface of the fiber, leading to temperature-induced phase separation, and the solvent-rich phase leads to porous structures. The solvent vapor pressure, solubility parameters of polymers and spinning solvents, polymer–solvent interactions, etc., influence pore formation. Further, condensation of water from highly humid atmosphere provides imprints on the fiber surface. The evaporation of condensed water gives porosity (vapor-induced phase separation). The use of a two-component system for making fibers followed by leaching/pyrolysis of one of the components can also provide porous fibers (Figure 2.26). The leachable materials could be a water-soluble polymer or salts for making a hydrophobic porous fiber. Here, one should simply use the difference in solubility of the two components for removal of one of them after fiber formation [44]. Low thermally degradable polymers can also be used as porogens (pore-forming material) in highly thermally stable fiber-forming material.

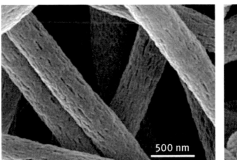 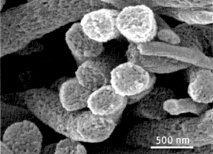

Figure 2.26: Polyacrylonitrile (PAN) porous fibers: surface morphology and cross section. The porosity is generated by the spinning of PAN and poly(vinyl pyrrolidone) (PVP). After fiber formation, the PVP was removed by extraction with water. [Reprinted with permission from *Macromol. Mater. Eng.* **2009**, *294* (10), 673–678, 2009. Copyright WILEY-VCH Verlag GmbH & Co. KGaA, Weinheim (2009).]

References

[1]   S.-C. Wong, A. Baji, S. Leng, *Polymer* **2008**, *49*, 4713–4722.
[2]   S. Agarwal, J. H. Wendorff, A. Greiner, *Advanced Materials (Deerfield Beach, Fla.)* **2009**, *21*, 3343–3351.
[3]   F.-L. Zhou, R.-H. Gong, *Polym. Int.* **2008**, *57*, 837–845.
[4]   A. Barrero, I. G. Loscertales, *Annu. Rev. Fluid Mech.* **2007**, *39*, 89–106.
[5]   L. S. Pinchuk, *Melt blowing: Equipment, technology, and polymer fibrous materials*, Springer, Berlin, New York, **2002**.

[6]   K. Nakata, K. Fujii, Y. Ohkoshi, Y. Gotoh, M. Nagura, M. Numata and M. Kamiyama, *Macromol. Rapid Commun.* **2007**, *28*, 792–795.

[7]   E. S. Medeiros, G. M. Glenn, A. P. Klamczynski, W. J. Orts, L. H. C. Mattoso, *J. Appl. Polym. Sci.* **2009**, *113*, 2322–2330.

[8]   (a) D. H. Reneker, A. L. Yarin, H. Fong, S. Koombhonse, *J. Appl. Phys.* **2000**, *87(9)*, 4531–4547; (b) A. Greiner, J. H. Wendorff, *Angewandte Chemie (International ed. in English)* **2007**, *46*, 5670–5703.

[9]   S. Agarwal, A. Greiner, J. H. Wendorff, *Prog. Polym. Sci.* **2013**, *38*, 963–991.

[10] O. O. Dosunmu, G. G. Chase, W. Kataphinan, D. H. Reneker, *Nanotechnology* **2006**, *17*, 1123–1127.

[11] X. Wang, H. Niu, T. Lin, X. Wang, *Polym. Eng. Sci.* **2009**, *49*, 1582–1586.

[12] B. Lu, Y. Wang, Y. Liu, H. Duan, J. Zhou, Z. Zhang, Y. Wang, X. Li, W. Wang, W. Lan et al., *Small (Weinheim an der Bergstrasse, Germany)* **2010**, *6*, 1612–1616.

[13] A. L. Yarin, E. Zussman, *Polymer* **2004**, *45*, 2977–2980.

[14] D. Wu, X. Huang, X. Lai, D. Sun, L. Lin, *J. Nanosci. Nanotech.* **2010**, *10*, 4221–4226.

[15] H. Niu, T. Lin, *J. Nanomater.* **2012**, 1–13.

[16] Z. Sun, E. Zussman, A. L. Yarin, J. H. Wendorff, A. Greiner, *Adv. Mater.* **2003**, *15*, 1929–1932.

[17] S. Jiang, G. Duan, E. Zussman, A. Greiner, S. Agarwal, *ACS Appl. Mater. Interfaces* **2014**, *6*, 5918–5923.

[18] V. Kalra, J. H. Lee, J. H. Park, M. Marquez, Y. L. Joo, *Small (Weinheim an der Bergstrasse, Germany)* **2009**, *5*, 2323–2332.

[19] P. Gupta, G. L. Wilkes, *Polymer* **2003**, *44*, 6353–6359.

[20] S. Chen, H. Hou, P. Hu, J. H. Wendorff, A. Greiner, S. Agarwal, *Macromol. Mater. Eng.* **2009**, *294*, 265–271; K. H. Roh, D. C. Martin, J. Lahann, *Nature Mater.* **2005**, *4*, 759.

[21] D. H. Reneker, A. L. Yarin, E. Zussman, H. Xu in *Advances in applied mechanics* (Eds.: H. Aref, Giessen, E. van der), Elsevier/Academic Press, Amsterdam, Boston, **2007**.

[22] Y. Filatov, A. Budyka, V. Kirichenko, *Electrospinning of micro- and nanofibers. Fundamentals and applications in separation and filtration processes*, Begell House, New York, **2007**.

[23] J. H. Wendorff, S. Agarwal, A. Greiner, *Electrospinning. Materials, processing, and applications*, John Wiley & Sons, Weinheim, Hoboken, NJ, **2012**.

[24] N. M. Neves, *Electrospinning for advanced biomedical applications and therapies*, Smithers Rapra, Shrewsbury, Shropshire, UK, **2012**.

[25] B. Ding, J. Yu, *Electrospun nanofibers for energy and environmental applications*, Springer-Verlag, Berlin Heidelberg, **2014**.

[26] M. Yousefzadeh, M. Latifi, M. Amani-Tehran, W.-E. Teo, S. Ramakrishna, *JEFF* **2012**, *7*, 17–23.

[27] S. Cai, H. Xu, Q. Jiang, Y. Yang, *Langmuir: ACS J Surfaces Colloids* **2013**, *29*, 2311–2318.

[28] L. H. Leung, S. Fan, H. E. Naguib, *J. Polym. Sci. B Polym. Phys.* **2012**, *50*, 242–249.

[29] M. Lee, H.-Y. Kim, *Langmuir: ACS J Surfaces Colloids* **2014**, *30*, 1210–1214.

[30] (a) Y. Si, J. Yu, X. Tang, J. Ge, B. Ding, *Nature Commun.* **2014**, *5*, 5802; (b) G. Duan, S. Jiang, V. Jérôme, J. H. Wendorff, A. Fathi, J. Uhm, V. Altstädt, M. Herling, J. Breu, R. Freitag et al., *Adv. Funct. Mater.* **2015**, *25*, 2850–2856.

[31] D. Zhang, J. Chang, *Nano Lett.* **2008**, *8*, 3283–3287.

[32] S. Pagliara, A. Camposeo, E. Mele, L. Persano, R. Cingolani, D. Pisignano, *Nanotechnology* **2010**, *21*, 215304.

[33] J. Doshi, D. H. Reneker, *J. Electrosta.* **1995**, *35*, 151–160.

[34] D. H. Reneker, I. Chun, *Nanotechnology* **1996**, *7*, 216–223.

[35] N. Tucker, J. J. Stanger, M. P. Staiger, H. Razzaq, Hofman, K, *J Eng. Fiber Fabr.* **2012**, *7 special issue*, 63–73.

[36] A. Bernet, M. Behr, H.-W. Schmidt, *Soft Matter* **2011**, *7*, 1058–1065.

[37] H. Jiang, M. O. Guler, S. I. Stupp, *Soft Matter* **2007**, *3*, 454–462.

[38] J. Shao, C. Chen, Y. Wang, X. Chen, C. Du, *Reactive and Functional Polymers* **2012**, *72*, 765–772.

[39] M. B. Sano, A. D. Rojas, P. Gatenholm, R. V. Davalos, *Ann. Biomed. Eng.* **2010**, *38*, 2475–2484.

[40] P. X. Ma, R. Zhang, *J. Biomed. Mater. Res.* **1999**, *46*, 60–72.

[41] M. Bognitzki, H. Hou, M. Ishaque, T. Frese, M. Hellwig, C. Schwarte, A. Schaper, J. H. Wendorff, A. Greiner, *Adv. Mater.* **2000**, *12*, 637–640.

[42] Z. Liu, D. D. Sun, P. Guo, J. O. Leckie, *Nano Lett.* **2007**, *7*, 1081–1085.

[43] M. Lallave, J. Bedia, R. Ruiz-Rosas, J. Rodríguez-Mirasol, T. Cordero, J. C. Otero, M. Marquez, A. Barrero, I. G. Loscertales, *Adv. Mater.* **2007**, *19*, 4292–4296.

[44] A. Gupta, C. D. Saquing, M. Afshari, A. E. Tonelli, S. A. Khan, R. Kotek, *Macromolecules* **2009**, *42*, 709–715.

3 Characterization of spinning solutions and fibers

3.1 Characterization of the spinning solution

3.1.1 Solubility

When electrospinning is carried out from a solution, one has to think about the right choice of solvent for the specific polymer. Besides other values such as electrical conductivity or vapor pressure of the solution, the solubility is the first factor to start the quest. Although for most common polymers several solvent systems are already reported for electrospinning, it is good to know which other options for the choice of solvent is possible, when the spinnability is not as good as desired using a particular solvent, when the hazard of the solvent is too high, or if the costs are disadvantageous. The polymer dissolution, in general, is a two-step process that involves solvent diffusion and macromolecular chain disentanglement. Therefore, the process is affected by thermodynamic interactions and diffusion of solvent. Some of the important parameters affecting solvent diffusion and chain disentanglements are temperature, crystallinity, polymer architecture, molar mass and molar mass dispersity, chemical structure, and composition in case of copolymers and stereochemistry. Because electrospinning from polymer solutions is highly affected by the choice of the solvent, the relevant parameters will help in finding the right system. For example, high-molar-mass polymers have high chain entanglements and therefore show decreased dissolution rates whereas high dispersities in molar mass help in having fast dissolutions. Chemical structure, composition, and stereoregularity can have different thermodynamic compatibilities with a particular solvent, providing different dissolution rates. A very good reading material for the physics of polymer dissolution is a review by Miller-Chou and Koenig [1].

Sometimes additional additives such as plasticizers can affect the solubility of the polymer in a particular solvent. A thermodynamically poor solvent (TPS) (i.e., nonsolvent for a polymer) can also act as plasticizer. In one of the studies, the addition of a definite amount of a TPS could increase the dissolution rate of a polymer in a thermodynamically good solvent (TGS). Hereby, TPS acted as plasticizer and increased the diffusion coefficient of the TGS and provided faster dissolution. For example, methanol and water are nonsolvents for poly(methyl methacrylate) (PMMA), but the addition of methanol and water in small amounts (till about 20 wt% for methanol and 6 wt% for water) doubles the dissolution rate of PMMA in methyl ethyl ketone (MEK) (Figure 3.1). This information could be helpful in preparing polymer solutions for electrospinning [2].

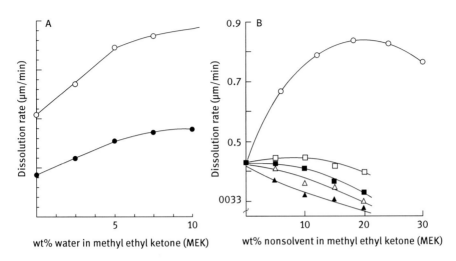

Figure 3.1: Dissolution rates of poly(methyl methacrylate) (PMMA) in methyl ethyl ketone (MEK) at 27.5 °C in presence of various amounts of (A) water [M_n (number average molar mass) = 320 × 10³ g/mol (•) and 36 × 10³ g/mol (O)] and (B) alcohols (O methanol, □ ethanol, ■ 2-propanol, △ 1-propanol, ▲ ethylene glycol). [Reprinted with permission from *J. Appl. Polym. Sci.* **1986**, *31*, 65–73, copyright John Wiley & Sons, Inc (1986)].

Another important point affecting polymer solubility is the crystallinity of polymers. Thermodynamically, the dissolution process for an amorphous polymer is governed by the free energy ΔG. If it is less than zero, the dissolution will happen spontaneously (second law of thermodynamics) (Equation 3.1).

$$\Delta G = \Delta H - T\Delta S \qquad (3.1)$$

ΔH is the change in enthalpy (energy content), T is the absolute temperature (in Kelvin scale) and ΔS is the change of entropy (degree of disorder) in Equation 3.1.

The change in entropy will be higher if the molecules are small. This means the polymers with lower degrees of polymerization (low molecular weight) will dissolve more easily than high-molecular-weight polymers. Also, a change in entropy for a dissolution process using solvents of small size will be more favorable for the dissolution process. This is reflected, for example, in a faster solubility of PMMA in solvents like tetrahydrofuran (THF) and methyl acetate (MA) than in bulky solvents like methyl isobutyl ketone (MIBK) [3]. This has also a kinetic basis as the diffusional coefficient is size dependent and is increased for smaller molecules [4].

In cases of crystalline polymers, the interactions between the molecules are very favorable, whereby the enthalpy of mixing becomes positive, which is disadvantageous for the dissolution. Furthermore, the intramolecular distances in crystal lattices are very small, making it difficult for a solvent molecule to penetrate. Many commonly used polymers for electrospinning such as polyamides (nylon-6, nylon-6,6) and polyesters [polycaprolactone (PCL), poly(L-lactide)] are semicrystalline and might require longer dissolution times and specific solvents for dissolution. Such polymers are soluble below their melting points only in solvents that have specific interactions with them. For example, highly crystalline polymers such as nylons are soluble at room temperature in solvents like phenol and formic acid due to the interactions between the polymers and solvent of hydrogen-bond type, whereas hydrocarbon semicrystalline polymers like polyethylene and polypropylene cannot be dissolved at room temperature due to the absence of any functional group capable of having interactions with the solvents. The easy solubility of polymers having functional groups such as amine, hydroxyl, carboxyl, amide and esters in appropriate solvents is due to hydrogen bonding between the polymer and the solvent.

For all high-molecular-weight polymers, generally used for electrospinning, the small positive entropy change and negative enthalpy of dissolution at standard temperature and pressure gives negative Gibbs free energy, making the dissolution process energetically favorable. In special cases, the solubility of a polymer does not increase with increasing dissolution temperature. Here, polymer–polymer or polymer–solvent interactions play an important role. Such polymers show a solubility gap in temperature – composition phase diagrams. This means that in some polymer–solvent systems an increase of the temperature is accompanied by insolubility. A well-known system that shows this effect is poly(N-isopropylacrylamide) (PNIPAm) in water, with a lower critical solution temperature (LCST) of ca. 32 °C. In this example, the hydrogen bonds between polymer and water break up around this specific temperature and the polymer precipitates. The contrary effect is also possible and is called the upper critical solution temperature (UCST), in which the polymer starts to dissolve over a specific temperature. A combination of LCST, UCST and more complex dissolution behaviors can also be expected (Figure 3.2) [5]. Temperature-dependent phase separation of polymers is known not only in water but also in various organic solvents. For example, polystyrene shows UCST behavior in cyclohexane and LCST in butylacetate. PMMA is soluble in many organic solvents like tetrahydrofuran, DMF and chloroform at any temperature between 0 and the boiling point of the corresponding solvents but shows LCST in acetone and UCST in acetonitrile. Before an electrospinning trial, one should check the literature of the polymer–solvent combinations or make experiments regarding solubility change with temperature, to bypass any surprises.

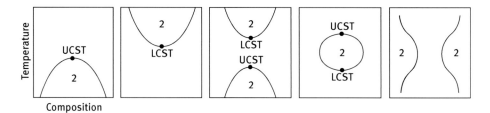

Figure 3.2: Miscibility gaps of LCST, UCST and mixed type; 2 represents two-phase region. [Reprinted with permission from *Macromol. Rapid Commun.* **2012**, *33* (22), 1898–1920. Copyright WILEY-VCH Verlag GmbH & Co. KGaA, Weinheim (2012).]

It would be good if the solubility of polymers in a particular solvent could be calcu-lated theoretically. This would help in choosing the correct solvent for electrospin-ning for any new polymer. Unfortunately, the solubility of a polymer in a solvent cannot be calculated so easily. Nevertheless, predictions can be done by using solu-bility parameters, which follow the principle of "like dissolves like" and put this approach into numbers. The first attempt was done by Hildebrand and Scott [6] by using the square root of the cohesive energy density E (Equation 3.2). This energy is needed to separate a mole of molecules of a solvent from each other and is a measure of intramolecular interactions.

$$\delta_\mathrm{H} = \sqrt{\frac{\Delta H_\mathrm{v} - RT}{V_\mathrm{m}}} \tag{3.2}$$

δ_H Hildebrand solubility parameter (MPa$^{\frac{1}{2}}$) or (($\frac{\mathrm{cal}}{\mathrm{cm}^3}$)$^{\frac{1}{2}}$); 1 MPa$^{\frac{1}{2}}$ $\approx$ 2.046 $(\frac{\mathrm{cal}}{\mathrm{cm}^3})^{\frac{1}{2}}$
R ideal gas constant (8.314 J/mol K)
T absolute temperature (K)
V_m molar volume (m^3/mol)
ΔH_v enthalpy of vaporization

Further, the solubility parameters are correlated to the enthalpy of mixing per unit volume for a binary mixture, which mainly governs the solubility process for macro-molecules (Equation 3.3). Since the enthalpy of mixing should be smaller than the change in entropy according to Gibbs free energy equation (3.1) for a dissolution process to occur, the solubility parameters δ for the two components should be very close to each other for a dissolution of a polymer in a particular solvent. If the differ-ence is less or equal to 3.7 MPa$^{\frac{1}{2}}$ the polymer should dissolve, which is a good guide-line for first solvent considerations in electrospinning approaches [7].

$$\frac{\Delta H_\mathrm{m}}{V} = (\delta_1 - \delta_2) \cdot \varphi_1 \cdot \varphi_2 \tag{3.3}$$

Hildebrand parameters are listed in literature for almost all known solvent and polymer types, so it is a very easy way to determine the right solvent for an electrospinning experiment. However, one should never forget that this is only an approximation and empirical tests should follow. Especially systems with polar interactions, hydrogen bond interactions and crystalline polymers often disagree with the Hildebrand parameters.

Hansen [7] refined the approach of Hildebrand by taking into account addition-ally the polar and hydrogen-bond interactions and consequently defined three solu-bility parameters as follows:

$$\delta_\mathrm{H} = \sqrt{\delta_d^2 + \delta_p^2 + \delta_\mathrm{h}^2} \tag{3.4}$$

δ_{H} Hildebrand solubility parameter

δ_{d} a value for the dispersion forces: London dispersion forces are higher with increasing amounts of free electrons

δ_{p} a value for the permanent dipolar intramolecular forces: dipole–dipole interactions

δ_{h} a value for the energy of intramolecular hydrogen bonds: electronic transfer

Similar to the Hildebrand parameters the Hansen parameters (HSP) have the unit of $(\mathrm{MPa}^{\frac{1}{2}})$ and can even be counted back (Equation 3.4). So the Hansen approach is a more defined way for searching the right solvent for electrospinning.

The calculation of solubility parameters for high-molar-mass polymers is not as straightforward as for low-molar-mass molecules since the enthalpy of vaporization of polymers is not measurable. The group contribution method is one of the most commonly used methods for the determination of solubility parameters of polymers. The dilute solution intrinsic viscosity of the polymer will also be maximum in a solvent that matches with it in solubility parameter. Therefore, the determination of intrinsic viscosity (η) in different solvents can also provide indirect measure of the solubility parameter of the polymer. The intrinsic viscosity can be measured in a laboratory using Ubbelohde capillary viscometer. The polymer solution viscosity (for neutral polymers) at any concentration is always more than that of pure solvent and the relative increase in the solution viscosity with respect to the pure solvent is represented by relative viscosity (η_{rel}). The intrinsic viscosity is defined by Equation (3.5).

$$[\eta] = \lim_{c \to 0} \frac{\eta_{\mathrm{rel}} - 1}{C} \tag{3.5}$$

The relative viscosities are measured for polymer solutions of different concentrations. A plot of $(\eta_{\mathrm{rel}} - 1)/C$ vs. polymer concentration (C) is a straight line with positive slope according to the Huggins equation (3.6). The extrapolation to the zero concentration gives intrinsic viscosity as the intercept. For polyelectrolytes, viscosity decreases with concentration and a plot of $(\eta_{\mathrm{rel}} - 1)/C$ vs. polymer concentration gives a straight line with a negative slope.

$$\frac{\eta_{\mathrm{rel}} - 1}{C} = [\eta] + K_{\mathrm{H}}[\eta]^2 C \tag{3.6}$$

Furthermore, for judging the solubility of polymers, 2D and 3D graphing systems were developed based on the three HSP, which can help by making an overview about the possible solvents to be used in electrospinning systems. One of these 2D graphing systems was shown by Teas. The 2D approach is often clearer and can be accomplished by plotting δ_{d} vs. δ_{p}, δ_{p} vs. δ_{h} or δ_{d} vs. δ_{h}. In order to illustrate all three components in one 2D graph, the Teas [8] approach can be adduced using fractional cohesion parameters $(f_{\mathrm{d}}, f_{\mathrm{p}}$ and $f_{\mathrm{h}})$. Thereby, an assumption is made to normalize the HSP in which the sum of the three parts amounts to 100% (or 1).

$$f_d = \frac{\delta_d(\cdot 100)}{\delta_d + \delta_p + \delta_h} \tag{3.7}$$

$$f_p = \frac{\delta_p(\cdot 100)}{\delta_d + \delta_p + \delta_h} \tag{3.8}$$

$$f_h = \frac{\delta_h(\cdot 100)}{\delta_d + \delta_p + \delta_h} \tag{3.9}$$

$$f_d + f_p + f_h = 1(\text{or } 100) \tag{3.10}$$

f_d, f_d, f_d normalized values for the HSP ($\delta_d, \delta_p, \delta_h$)

The fractional parameters are drawn as three sides of a triangle, and each solvent gets a unique position in the Teas graph depending upon its polarity, hydrogen-bonding capacity, etc. This ternary solubility diagram could be highly useful in selecting appropriate solvents or solvent mixtures for electrospinning, and the solubility position of the particular polymer can be described in this diagram.

    The proper choice of solvent for electrospinning is not only highly important for getting a stable spinning process providing smooth fibers but can also influence the physical properties of fibers. Luo et al. [9] have studied the influence of 49 solvents of different solubility parameters on the electrospinnability of polycaprolactone (PCL) and mapped it in the form of a Teas plot (Figure 3.3). The solvents showing very good solubility (immediate dissolution) such as cyclohexanone, acetophenone, benzene, toluene, bromobenzene and styrene were not appropriate for spinning and no fibers were obtained. Solvents like chloroform, THF and dichloromethane were suitable for getting fibers. This work, combined with previous works by the authors, shows that solvents with a high solubility are not always good for electrospinning. Solvents in which polymers were partially soluble were superior in electrospinning in comparison to the good solvents. Besides that, Teas graph can also be used for selecting solvents of different solubility and spinnability for making binary solvent systems to get a stable and reproducible electrospinning system. This could help in using less-toxic, nonflammable solvents that otherwise were not suitable alone for electrospinning of a particular polymer. The knowledge about the solubility of a particular polymer in different solvents can also be utilized in producing porous fibers by choosing appropriate combinations of solvent–nonsolvent mixtures (not necessarily high vapor pressure), leading to liquid–liquid phase separation in electrospun fibers [10].

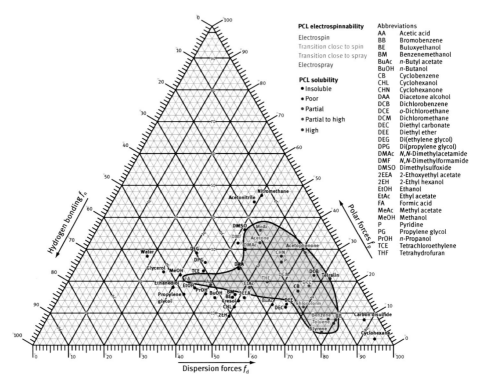

3.1.2 Electrical conductivity

Especially in the case of electrospinning, a certain conductivity of the spinning solution is required and can be seen as the underlying force of a solution to elongate in the direction of a collector with a converse charge (closing of the circuit, electron transfer). The electrical conductivity is a constant of proportionality for the current density and the electric field strength and is the reciprocal of the specific electrical resistivity. Therefore, it is a factor for how much a material is willing to conduct charges.

$$J = \sigma \cdot E \tag{3.11}$$

J current density (A/m²)
σ electric conductivity (S/m) or (1/Ω m)
E electric field strength (V/m)

In general, it cannot be said which conductivity is needed for a proper electrospinning system but the characteristic conductivity of spinning solutions from literature is 0.05–30 mS/m. However, with increasing surface tension, a higher conductivity is needed as counterforce. With higher conductivities, lower polymer concentrations can be used for the spinning process. Since lower polymer concentrations lead again to smaller diameters, an increase in the conductivity is a good aid for reducing fiber diameters or bead formations [11, 12]. The same applies to the viscosity of the spinning solution. Sometimes, the fiber-forming weight concentration of a polymer is too high to be still in an adequate viscosity range required for getting thin fibers. With the help of conducting salts, the required weight concentration can be reduced, whereby the viscosity decreases (see Section 3.1.3).

Almost all kinds of inorganic ionic salts can be used for increase of the conductivity of the spinning solution, provided that they are soluble in the particular solvent. Some commonly used inorganic and organic salts in electrospinning formulations are for example NaCl, NaHCO$_3$, CaCl$_2$, MgCl$_2$, NaNO$_3$, KCl, NaH$_2$PO$_4$, or organic salts such as benzyltriethylammonium chloride (TEBAC), tetra-n-butylammonium bromide (TBAB) or pyridinium formate (PF). The use of volatile salts such as PF for modulating the conductivity of the spinning solution is advantageous due to its easy removal from the fibers simply by heating at high temperature in vacuum (pay attention not to exceed the glass transition temperature in case of amorphous polymers, so the fiber morphology will not change).

The typical way to measure the conductivity of electrospinning solutions is to use a conductivity meter. It consists of minimum two electrodes with a known surface and a defined distance. Between those electrodes, the resistance is measured. Since the potentials of electrical bilayers can alter the result, porous platinum electrodes are used, whereby the risk of false measurement is reduced. However, they have the disadvantageous necessity of being stored under liquid, in order to protect the electrodes from ruptures due to drying processes.

A reliable measurement always requires calibration of the measuring system with a standard solution, which is usually a KCl solution with an exact conductivity, in the range of milli- to microsiemens per centimeter. Since the conductivity is temperature dependent, the measurement is carried out at the same temperature (usually 25 °C, the temperature for the specific conductivity of the standard solution is defined). After the meter is calibrated, the probe head should be washed thoroughly with the solvent of the electrospinning solution and immersed in the spinning solution for conductivity measurement (be aware of the minimum immersion depth). Once the value for the conductivity is stable, it can be recorded. At the end of the tests, the washed electrodes should be stored again under the appropriate storing solution.

3.1.3 Rheology

3.1.3.1 Theoretical background

Rheology is the study that addresses the flow of matter. Since the electrospinning process utilizes polymer solutions or polymer melts, one can determine the rheological behavior before carrying out electrospinning. Furthermore, it is a good method for the determination of the impact of particle sizes, polydispersities, or volume fractions of multi component systems on the flow behavior. With these kinds of measurements it is possible to determine whether an electrospinning solution is stable over long time period, a solution or dispersion is homogenous, or a spinning solution is in a viscosity range, where an electrospinning experiment is possible.

The polymer chain entanglements decide the rheological behavior. In dilute solutions, the macromolecular chain entanglement decreases and reaches a point on further dilution where there are no entanglements, which are not very favorable for getting fibers by electrospinning. In this situation, the polymer solution has a so-called Newtonian behavior and the following equation is applied:

$$\tau = \eta \cdot \dot{\gamma} \tag{3.12}$$

τ shear stress (Pa)
η viscosity (Pa s)
$\dot{\gamma}$ shear rate (s^{-1})

Newtonian behavior is also observed for entangled polymer solutions (pseudoplastic solutions) at very low shear rates and the corresponding solution viscosity (zero shear viscosity: η_0) is characterized for viscous spinning solution (Figure 3.4). At low shear rates, the density of entanglement stays the same, since they can be reformed. An increase in the shear rate leads to more nonrecoverable disentanglements. This situation leads to shear thinning. The specific moment, at which the shear thinning starts (the minimum time for the entanglement to fully recover) is called the characteristic retardation time Λ_0 and is defined as the inverse value of the shear rate at that point (Figure 3.4). The Newtonian behavior is also seen at very high shear rates where the alignment of the polymer chains is at its maximum, providing a second Newtonian region (infinite shear viscosity: η_∞). During the elongation of the jet in electrospinning the shear rates are very high (in ranges of about 10^4 s^{-1}) which makes highly viscous solutions spinnable and could lead to polymer chain alignment.

Besides pseudoplastic fluids, the shear rate dependent bearings are also valid for dilatant fluids (starch suspension, dispersions with repulsive interactions) and time-dependent effectuations such as thixotropic (ketchup) and rheopectic fluids (gypsum pastes). A comparison of the shear rate dependent viscosity and shear stress behavior for pseudoplastics, Newtonian and dilatant fluids is shown in Figures 3.5 and

3.6, respectively. The thixotropic and rheopectic solutions are not typical for electrospinning solutions and hence they will not be discussed in this book.

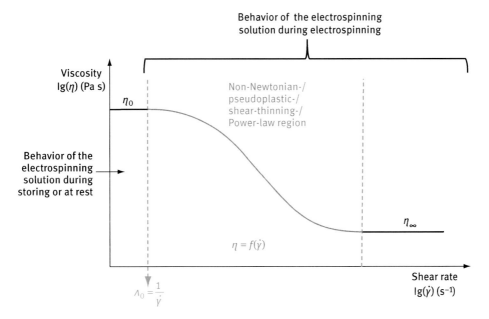

Figure 3.4: Pathway of the viscosity for entangled polymer solutions and melts with ascending shear rates.

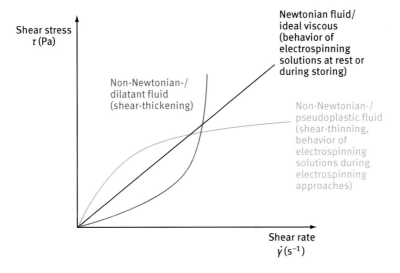

Figure 3.5: Exemplary behaviors of the shear stress with ascending shear rates.

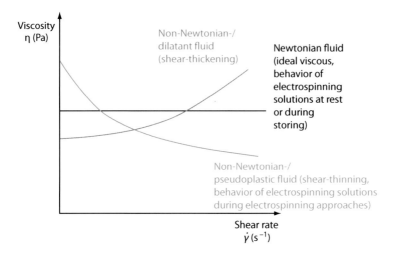

Figure 3.6: Exemplary behaviors of the viscosity with ascending shear rates.

A typical non-Newtonian behavior of viscoelastic fluids (such as polymer solutions for electrospinning) is the occurrence of the "Weissenberg effect" (Figure 3.7). The rod twisted inside of a Newtonian fluid exhibits a deepening of the solution, whereas a concentrated spinnable polymer solution winds-up.

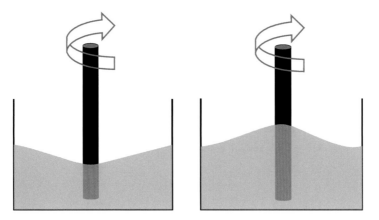

Figure 3.7: Influence of the Weissenberg effect on a Newtonian fluid (*left*) and a non-Newtonian (electrospinning solution), viscoelastic fluid (*right*).

A big benefit of rheology is the insight in molecular motions. Shorter polymer chains react differently under applied stress than longer chains do and the same is true for higher and lower polymer concentrations in solutions because of the already mentioned variable entanglements. Hence it is possible to determine when the entanglements in polymer solutions or melts start to form. These entanglements are needed in electrospinning approaches, since they ensure the fiber-forming ability of a polymer solution [13]. Higher entanglements can be achieved in two different ways: by increasing the polymer concentration in a polymer solution or by using polymers with higher degrees of polymerization, i.e., higher molar mass polymers (Figures 3.8 and 3.9).

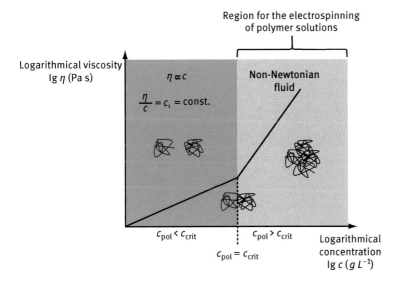

Figure 3.8: Dependence of shear viscosity and concentration in polymer solutions with constant molecular weights. C_{pol} = polymer concentration; C_{crit} = critical polymer concentration.

The lowest limiting value of the viscosity in pseudoplastic (viscoelastic) fluids (e.g., polymer melts and solutions) is the zero shear viscosity. A polymer solution without any entanglements just shows a Newtonian behavior (no dependence of shear rate and viscosity), so there is only one constant shear viscosity to be addressed. Below the critical polymer concentration (C_{crit}) (the entanglements are completely built up at C_{crit}) the viscosity and concentration are related in a direct proportionality in a good solvent, followed by a steep increase of the viscosity with increasing concentrations above C_{crit} due to extensive chain entanglements. The same applies to the molecular weight, i.e., molecular weight increases with viscosity ($\eta_0 = M^\alpha$), α is approximated with a value of 3.4 and fits well for not cross-linked polymers. For high-molecular-weight polymer solutions, the characteristic retardation time (Λ_0) is increased. This means that, with higher molecular weights, lower polymer concentrations for electrospinning can be used.

The concentrations and molecular weights of polymers used in electrospinning approaches should be higher than the critical molecular weight and critical concentration, so that the polymer chains have already a strong entanglement when being

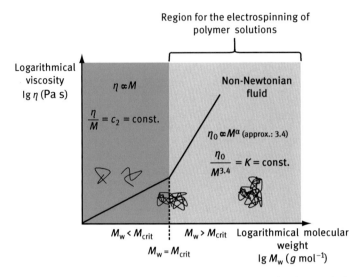

Figure 3.9: Dependence of shear viscosity and average molecular weight in polymer solutions and melts.

ejected through a nozzle [13]. In other words, the solution should show viscoelastic properties. Besides that, the viscosity of spinning solutions with high-vapor-pressure solvents should be low (Figure 3.10), otherwise the solution runs dry before a jet is formed and only relatively big particles are sprayed onto the collector. An overview about different shear viscosity values is given in Table 3.1. Hence the region for a good electrospinning can be determined either by a trial and error process or by testing the rheology of a spinning solution.

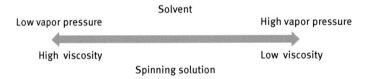

Figure 3.10: Relation between vapor pressure of the solvent and viscosity of the spinning solution. Low viscosities of the polymer solutions (lower polymer concentrations or molecular weights) can be used with high-vapor-pressure solvents, whereas low-vapor-pressure solvents need higher viscosities of the polymer solution (higher polymer concentrations or molecular weights).

Table 3.1: Zero shear viscosities of different electrospinnable polymer solutions. Water and honey shear viscosity values are shown for comparison.

Substance	Molecular weight (g/mol)	Concentration (wt%)	Solvent	Shear viscosity/zero shear viscosity (mPa•s)
Water				1
Honey				10^4
Polymer melts				10^3–10^6
PET-co-PEI	$M_w = 77 \times 10^3$		$CHCl_3$/DMF 70/30 wt/wt	<30 [14]
PCL	$M_w = 50 \times 10^3$	11	glacial acetic acid + 2wt% TEA	10^2 [15]
PVB	$M_w = 60 \times 10^3$	10	EtOH	100 [16]
PVB	$M_w = 60 \times 10^3$	10	MeOH	60 [16]
PA6	$M_w = 20 \times 10^3$	34	Formic acid	1.2×10^3 [17]
PAN	$M_v = 14 \times 10^4$	7	DMF	338 [18]
PAAc	$M_v = 45 \times 10^4$	8	Water	247 [19]
PAAc	$M_v = 45 \times 10^4$	8	DMF	3,913 [19]
PEO	$M_v = 4 \times 10^5$	7	Water	10^4 [20]
PVOH	$M_w = 146,000$– 186,000; 87–89% hydrolyzed	7.6	Water	738

PET-co-PEI: poly(ethylene terephthalate-co-ethylene isophthalate); PCL: polycaprolactone; PVB: poly(vinyl butyral); PA6: nylon-6 or polycaprolactam; PAAc: poly(acrylic acid); PAN: polyacrylonitrile; PEO: polyethylene oxide; PVOH: poly(vinyl alcohol). M_v = viscosity avearge molar mass; M_w = weight average molar mass.

3.1.3.2 Typical measuring systems for rheological behaviors

There is a variety of techniques for quantification of the rheological characteristics of solutions or dispersions, such as flow cups (the time of flow of a fluid is quantified for a cup with a defined orifice and volume), falling ball viscometer (the time required by a ball for moving a definite distance in a fluid is determined), or rotational rheometer. For the latter, different types of measurement geometries can be used, nevertheless the working principle is always the same: Apply torque and measure deformation (in a transferred sense: shear rate) or apply deformation and measure torque.

Cone and plate measuring system

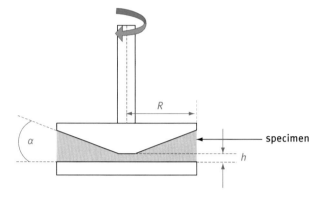

specimen

Figure 3.11: Cone and plate geometry.

$$\tau = \frac{3M}{2\pi R^3} \tag{3.13}$$

τ   shear stress (Pa)
M torsional moment (Nm)
R radius (m)

$$\dot{\gamma} = \frac{\dot{\varphi}}{\tan \alpha} \tag{3.14}$$

$\dot{\gamma}$ shear rate (s^{-1})
$\dot{\varphi}$ angular velocity (rad/s)
α   opening angle of the cone (rad)

The advantages in using the cone and plate geometry (Figure 3.11) are reasoned by a homogeneous distribution of the deformation and the shear rate in the whole sample. Furthermore, the required amount of the specimen is low. However, the gap height h (often around 50 µm for an opening angle of 1°) is unfavorable in cases of dispersions. Therefore, the maximum particle sizes in dispersions should not exceed a tenth of the gap height if this method is being used.

Parallel-plate measuring system

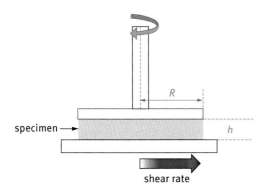

Figure 3.12: Parallel-plate geometry.

$$\dot{\gamma}_R = \frac{R \cdot \dot{\varphi}}{h} \qquad (3.15)$$

$\dot{\gamma}_R$ marginal shear rate (s^{-1})
h gap height

$$\tau_{R,N} = \frac{2M}{\pi R^3} \qquad (3.16)$$

$\tau_{R,N}$ marginal shear stress for Newtonian behavior (Pa s)

Parallel-plate geometry is usable for a rheological characterization of polymer solutions and dispersions. This geometry is more favored for polymeric dispersions as there is no restriction to the gap height (h) and polymeric dispersions with even large sized particles can also be measured. In cases of parallel-plate geometries (Figure 3.12) the shear rate rises continuously with the radius (R) because of the constant gap height h. Therefore, the maximum value for the shear rate is located at the outside margin, which is used for analyses. Sometimes the average shear rate, which is two-thirds the value of the marginal shear rate, is also used for analysis.

Rotational experiments

In rotational experiments, either the shear stress or the shear rate can be predefined (controlled shear stress test or controlled shear rate test). To achieve the viscosity function and thereby the zero shear viscosity, the controlled shear rate test is carried out as it can be seen in Figures 3.4 and 3.6. The shear rate can be adjusted either in a linear or stepwise fashion with time. The latter should be favored in cases of viscoelastic fluids in order to minimize transient effects. The controlled shear stress test is a classical test for the determination of the yield point (e.g., of tooth pastes or other networking substances, which can be disadvantageous in cases of electrospinning).

In a controlled shear rate test, to find the best ranges for a new fluid, one has to play with the settings but a first approach could be as follows:
- Range of shear rates: 0–2,000 s^{-1}
- Amount of measuring points: 50
- Equilibration time for each measuring point: 6 s
- Testing time: 300 s

The Newtonian region of the zero shear viscosity is often below $\dot{\gamma} < 1$ s^{-1}, so a logarithmic setup is preferable. However, a testing instrument with aerodynamic bearings should be used. The guideline for a setting with aerodynamic bearings should be: the lower the shear rate, the higher the equilibration time:
- Range of shear rates: 0.01–2,000 s^{-1}
- Amount of measuring points: 50
- Equilibration time at the beginning (low shear rates): $\frac{1}{\dot{\gamma}}$, for example, 100 s for 0.01 s^{-1}
- Equilibration time for higher shear rates: 6 s

Oscillatory experiments:

In oscillatory experiments, there are transient shear conditions. That means every point of a measurement has different deformations or distinct shear stresses. Therefore, a complex shear modulus $G^\star$ and a complex viscosity $\eta^\star$ are defined as follows:

$$G^* = \frac{\tau(t)}{\gamma(t)} \tag{3.17}$$

G^* complex shear modulus (Pa)
$\tau(t)$ time-dependent shear stress (Pa)
$\gamma(t)$ time-dependent deformation (%) or (1)

$$\eta^* = \frac{\tau(t)}{\dot{\gamma}(t)} \tag{3.18}$$

η^* complex viscosity (Pa s)
$\dot{\gamma}(t)$ time-dependent shear rate (s^{-1})

$$|G^*| = \dot{\varphi} \cdot |\eta^*| \tag{3.19}$$

$\dot{\varphi}$ angular velocity (rad/s)

However, an empirically discovered rule after W. P. Cox and E. H. Merz [21] (Cox–Merz rule) implies congruent values for η and $\eta^\star$ as long as the shear rates and angular frequencies are matching.

$$\eta(\dot{\gamma}) = |\eta^* (\dot{\varphi})| \tag{3.20}$$

This rule takes effect for almost all unfilled polymer melts and solutions in the region of the zero shear viscosity and sometimes beyond that.

A benefit in dynamic test conditions is the possible cleavage of the elastic and the viscous signals of a specimen (how big is the viscous fraction in comparison to the elastic fraction of an electrospinning solution with ascending shear rates), which are called G' (storage modulus) and G'' (loss modulus), respectively. Concerning this, the preset value of the test (e.g., shear stress) is phase shifted to the response (e.g., deformation) by a loss angle δ. The different possible situations depending on the material characteristics could be:

-   Ideal elastic behavior (completely rigid substances): Shear stress and deformation oscillate in the same phase; shear rate is shifted by 90°
-   Ideal viscous behavior (Newtonian fluid): The oscillation of shear stress and deformation are phase shifted by $\delta = 90°$; shear stress and shear rate are in phase (Figure 3.13)
-   Viscoelastic behavior (e.g., polymer solutions): The oscillation of shear stress and deformation are phase shifted by $0° \leq \delta \leq 90°$

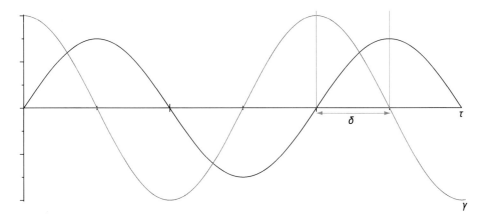

Figure 3.13: Phase shift of shear stress and deformation in ideal viscous fluids.

A variable that combines the loss and storage modulus is the dissipation factor tan(δ), which is consequently a factor to describe the sol–gel transition of a solution or melt.

$$\tan(\delta) = \frac{G''}{G'} \tag{3.21}$$

tan(δ)   Dissipation or loss or damping factor (arbitrary unit)

When tan(δ) is below 1, the specimen shows less viscous than elastic properties. This means that below a dissipation value of 1 an electrospinning approach will not function probably, since the behavior of the spinning solution is more like that of a solid and not like that of a liquid anymore.

For the testing of a new specimen, it is initially important to distinguish the linear viscoelastic region from the deformation. Therefore, an amplitude sweep is the right choice (e.g., preselection of the deformation with a constant frequency). The test setting could be applied as follows:
- Logarithmic preselection of the deformation: 0.01–100%
- Amount of measuring points: 10 points per decade
- Frequency: $f = 1\,Hz$; $\dot{\varphi} = 2\pi{\cdot}f$; $1\,Hz \triangleq 6.28\,rad\,s^{-1}$

Most of the time, the storage modulus G' is the first to decrease. That is why it is used for the definition of the linear viscoelastic region, as can be seen in Figure 3.14. The value at which the linear behavior of the storage modulus drifts is the limiting deformation γ_L of the linear viscoelastic region. Deformations with higher values will destroy the inner structure of the specimen, sometimes even in an irreversible matter.

Afterward, frequency sweep tests can be done with a constant deformation (value below the detected limiting deformation γ_L) and preselected frequencies (or angular velocities $\dot{\varphi} = 2\pi f$), to achieve the viscosity function and the storage and loss modulus function of the electrospinning solution, which could be as follows:
- Logarithmic preselection of the angular velocities or frequencies: 200–0.1 rad/s or 30–0.02 Hz
- Amount of measuring points: 10 points per decade
- Deformation: $\gamma \leq \gamma_L$, e.g., 1% or 0.01

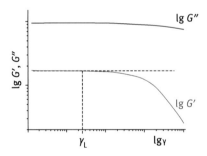

Figure 3.14: Amplitude sweep of a polymer solution with preselected deformations. Here, the loss modulus G'' is always higher than the storage modulus, hence the specimen has only a sol character in this measuring region. γ_L: limiting deformation of the linear viscoelastic region.

Due to the different shear rates in oscillatory experiments comparing to rotational tests (especially when no aerodynamic bearings are applied) the two measurements can complement each other. For a typical interpretation, the logarithmic angular velocities $\dot{\varphi}$ (or logarithmic frequencies f) are plotted on the x-axis, whereas the logarithmic loss G'' and storage moduli G' [alternative: the logarithmic dissipation factor $\tan(\delta)$] are plotted on one y-axis and the logarithmic complex viscosity $|\eta^*|$ on another y-axis (Figure 3.15). However, it is not reasonable to use the values of the complex viscosity if G' is higher than G'', since the Newtonian law (Equation 3.12) holds only for shear rates greater than zero (if G' is higher than G'', there is no shear rate, since the substance behaves like a solid, which is twisted/elongated;

shear = flow of matter). In such cases the dissipation factor or loss and storage moduli can be used for the evaluation.

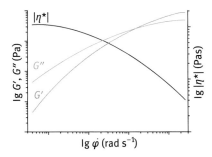

Figure 3.15: Result of a frequency sweep test on a polypropylene melt.

3.2 Characterization of the fibers

In order to characterize the fibers produced by electrospinning it is necessary to distinguish between the molecular structure (Angstrom to nanometer regime) and the resulting morphology of the fiber (nanometer to micrometer scale). For the latter, Section 3.2.1. (Morphology and size) is the right choice and the former topic can be found in Section 3.2.2 (Molecular structure).

3.2.1 Morphology and size

3.2.1.1 Optical microscopy

The optical microscopy is the easiest and fastest way to observe the obtained fibers after electrospinning. However, it is not the method of choice for quantitative measurement of fiber diameters, although it is the fastest way to follow fiber formation by electrospinning. For example, bead-like structures, electrospraying or fiber breakages can be seen with an optical microscope, whereas the pore structures of fibers are not measurable. For the latter, a scanning electron microscope (SEM) or an atomic force microscope would be the right choice. Therefore, it is good to know what limits the analysis of electrospun fibers with an optical microscope.

The critical term of an imaging technique to differentiate the observed objects (e.g., fibers), is the resolution and the contrast and not the magnification. These values are not only related to the quality of a device, they are rather connected to the wavelength of the detecting particles (e.g., photons). The maximum resolution is associated as the minimum distance Δx at which two objects can still be distinguished (e.g., if there is one or more fibers; Equation 3.22). So it is also the minimum of an object to be sharply seen:

$$\Delta x = \frac{0.61 \cdot \lambda}{n \cdot \sin(\alpha)} \tag{3.22}$$

The denominator of the fraction is the numerical aperture NA of E. Abbe [22], where α is the half field angle of the objective and n is the refractive index of the medium between object and objective (Equation 3.23). The value of the numerical aperture can typically be found on the objective, e.g., 100 × (magnification)/0.9 (numerical aperture) and can be seen as an index of the luminous efficiency for making detailed observations of fibers.

$$NA = n \cdot \sin(\alpha) \tag{3.23}$$

Since the wavelength of visible light is located around 380–780 nm, the maximum resolution under best conditions for a numerical aperture of 0.9 (already very high) is about 258–529 nm. For real conditions, this means that in dependence of the light source and the objective, structures (like the diameter of electrospun fibers) below ca. 600 nm cannot be seen sharply anymore and no quantitative measurements should be conducted. Nevertheless, it is possible to distinguish if the electrospinning process gives fibers or particles or if the achieved fibers have beads.

In order to obtain better resolutions it is possible to change the wavelength of the light source, but then the radiation needs to be translated into visible light. Therefore, a straightforward approach is to change the refractive index of the medium between the object and objective, which can be done via immersion oil and an oil immersion objective (Figure 3.16).

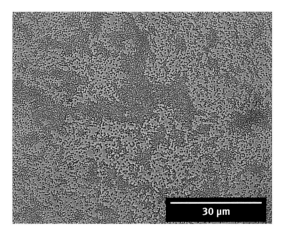

Figure 3.16: Optical microscope image of polystyrene (PS) particles via oil immersion technique (objective: 100×/1.25). The spheres are smaller than the resolution of the setup (ca. 200 nm), so that in denser regions they cannot be seen separately. However, it is possible to say that there are purely particles and no fibers.

Typical refractive indices of immersion liquids are about 1.5 (instead of 1 for air) [23] and numerical apertures above 1 can be reached. Nevertheless, a cover slip is needed with almost the same refractive index to prevent contacts between oil and specimen (e.g., electrospun fibers; Figure 3.17).

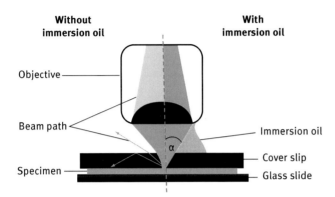

Figure 3.17: Schematic setup of microscopy with and without immersion oil. If immersion oil is used, more light (i.e., information) reaches the objective.

Another reason that objectives with higher magnifications have commonly a higher numerical aperture is owing to the closer distance between the objective and the sample. Hence one should start from a lower magnification to higher augmentations so as to prevent the objective from destroying the sample (or the other way around). Moreover, it is way easier to find the right focus.

In consequence of the small diameters that can be obtained by the electrospinning process, it is debatable if the measurement of the fiber diameter in optical microscope images is reliable (e.g., error is too high). Therefore, it is better to use a calibrated SEM.

It is also easier to observe only a few electrospun fibers instead of a nonwoven mat due to contrast and focus problems in optical microscopes. The latter can be overcome by *focus stacking* (Figure 3.18) but there is still the problem of the contrast, which makes it difficult especially for aligned fibers (fibers with small diameters lying parallel to each other, so the resolution is too low for differentiating the borders of the fibers). To gain a good observable specimen under an optical microscope, the modest way is to spin directly onto a glass slide for a few seconds (Figure 3.19; do not forget to ground the electrodes first, in order to prevent an electric shock).

Figure 3.18: Focus stacking on a fiber mat with aligned fibers. Only the perpendicularly oriented fibers give enough contrast to be poorly detected.

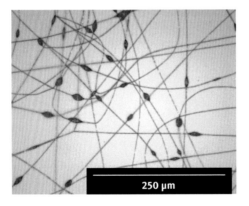

Figure 3.19: Optical microscope of electrospun polystyrene (PS) fibers on a glass slide. The concentration of the polymer solution was too low, so that beaded fibers were the consequence.

3.2.1.2 Electron microscopy

Electron microscopy is one of the frequently used methods for observing fiber morphology and quantifying the fiber diameters made by electrospinning. Therefore, in this section we will discuss it in some details.

In Section 3.2.1.1 (optical microscopy) the wavelength-dependent maximum resolution of an optical microscope was addressed. This extends to the resolution of other microscopy methods, too, because of the diffraction of irradiations. Hence in electron microscopes electrons are used as source of radiation. In contrast to a photon (no assumed mass), an electron has a given mass (m_e: 9.109×10^{-31} kg). According to L. de Borglie [24] relationship for an assigned acceleration voltage for an electron a distinctly smaller wavelength is achieved (about 10^4 to 10^5 times smaller for acceleration voltages of 1–200 kV) (Equation 3.24) and makes therefore the observation of fiber morphologies possible even with very small diameters.

$$\lambda = \frac{h}{\sqrt{2m_e \cdot e \cdot U}} \tag{3.24}$$

λ wavelength (m)
h Planck's constant (6.626×10^{-34} Js)
e elementary charge (1.602×10^{-19} C)
U acceleration voltage (V)

A particular drawback in using electrons as imaging source is the energy that is exerted to the specimen, because not all of the electrons are scattered elastically and one can easily damage a sample by heat. Precisely in the case of electrospun fibers, mostly organic materials are used, which are not very heat resistant and modified appearances could arise. When light is used for an imaging exploration, it is easy to interpret the results, because it is similar to what the brain does in everyday life. However, in the world of electrons, a different approach for contrast, colors, and shapes needs to be used. This will be discussed in the next sections regarding two different types of electron microscopes: the SEM (Figure 3.20) and the transmission electron microscope (TEM).

Scanning electron microscope

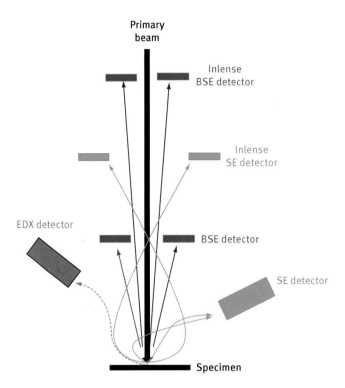

Figure 3.20: Possible detector setup of a scanning electron microscope. There can be one detector or more. The secondary electrons (SEs) are chiefly used for topographic imaging, whereas the material contrast can be achieved by back-scattered electrons. More information about the molecular composition of a substance can be obtained by characteristic X-rays via an energy-dispersive X-ray spectroscopy (EDX).

For a topographic analysis of a fiber mat a SEM is the right choice. As a result, it is the most crucial method for assaying the morphologies of the produced fibers by electrospinning. In SEM analysis, the electrons are screened dot-like over the specimen and the amount of the differently reflected and scattered electrons is quantified. The secondary electron (SE) detector is the most common detector used for SEM analysis of electrospun fibers. The inelastically scattered electrons are detected by SE. For composite fibers, i.e., metal nanoparticles–polymer fibers the SE detector would provide only the surface morphology but for seeing a better contrast/components separately, back-scattering detectors (BSE) are used. A BSE detector detects the elastically scattered electrons from the sample. The polymer fraction will appear dark, whereas the metal content will appear bright in SEM images using BSE detector. The resolution of the BSE detector is much less than that of the SE detector because the volume from which the BSE originates is bigger in comparison to that of SE (Figure 3.21). This diminishes the scope of resolution. Newer back-scattering techniques such as inlense BSE detectors can overcome the problem of low resolutions by higher efficiencies and lower acceleration voltages can be used, whereby the domain of excitation decreases.

If the same spot of the fiber mat is measured with a secondary electron detector (at lower voltages), a good comparison can be seen in the image qualities (Figure 3.22). Because of less energy contents (in the range of <50 eV in contrast to >50 eV to keV for BSE), the underlying stimulated electrons cannot reach the surface and only the upper nanometer range of the specimen is detected. Since typical SE detectors are located on one side of a sample, the image emerges as it has been taken from top with an illumination located on the side. Thus the fiber image arises with a high spatiality due to different kinds of contrast effects, such as shadowing effects or edge effects (Figures 3.22 and 3.23). By using an inlense SE detector the image has less spatiality, since the illumination seems to appear from overhead. However, lower working distances can be utilized, whereby higher amounts of secondary electrons can be collected and less acceleration voltage can be applied, resulting in increased potentials for the resolution for observing fiber morphologies with small diameters.

The selection of the matching acceleration voltage is dependent on the measuring system and the specimen, but in cases of fiber mats an acceleration voltage below 5 kV is suitable for secondary electron detectors. A back-scattering detector should be applied with acceleration voltages ≥10 kV.

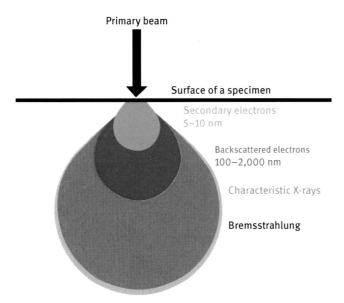

Figure 3.21: Schematic layout of the resolution for different kinds of electrons. Due to the distinct energy contents of the electrons, the interaction volumes range from few nanometers to several micrometers, influencing the resolution.

Figure 3.22: SEM image of electrospun poly(vinyl alcohol) (PVOH) fibers. (A) Acceleration voltage: 20 kV, BSE detector; (B) acceleration voltage: 20 kV, inlense detector; (C) acceleration voltage: 20 kV, SE detector; (D) acceleration voltage: 1 kV, mix of 25% SE inlense detector and 75% SE detector located at the side.

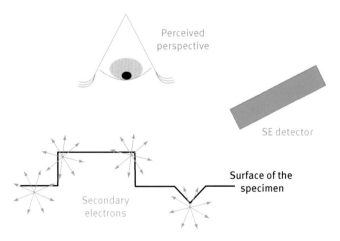

Figure 3.23: Development of an image, taken by a secondary electron detector. The more arrows (secondary electrons) escaping the surface in the direction of the SE detector, the brighter the position in an image.

Sample preparation

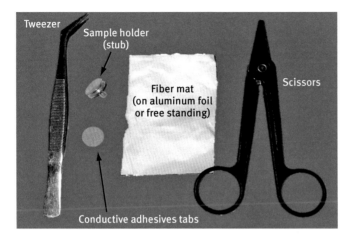

Figure 3.24: Required tools for sample preparation in SEM analysis of electrospun fiber mats.

The materials required for sample preparation are shown in Figure 3.24. Never touch the sample and sample holder without gloves in order to prevent fatty residues that can lower the image quality by volatilization inside of the electron microscope. For sample preparation follow the steps (Figure 3.25): (A) adhere the conductive adhesive tab onto the stub by using tweezers, (B) cut out a small piece of the electrospun fiber mat, (C) remove the covering layer of the conductive adhesive tab and place the piece of the fiber mat on top of it by using tweezers. Be careful not to touch too much of the fiber mat, for example, just hold at the edges if required. Assure that the fiber mat is fixed by pressing on the borders of the fiber mat. The specimen should be adhered properly, because a strong vacuum will be applied. The specimen is usually coated by a sputter in front of the measurement. For example, a 1.3 nm platinum layer is sufficient most of the time.

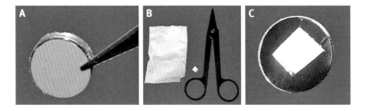

Figure 3.25: SEM sample preparation.

Transmission electron microscope

TEM is one of the most important methods for the analysis of blend or composite fibers, commenting about core-shell morphology, presence of metal nanoparticles in fibers, etc. The analysis of electrospun fibers with a TEM is only reasonable when different kinds of morphologies are included inside of a fiber, otherwise the fibers will appear transparent and no surface morphologies are detectable (Figure 3.26).

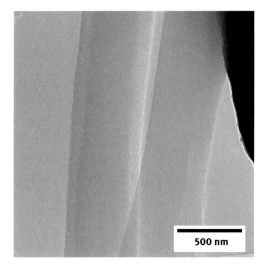

500 nm

Figure 3.26: TEM image of electrospun porous polystyrene (PS) fibers without filler. The fibers appear transparent, so no surface morphology can be measured.

In conventional TEMs a detector with spatial resolution is used and the analyst is directly able to see the image of the sample (e.g., of fibers) on a fluoroscopic screen or one can just use a photographic plate as detector.

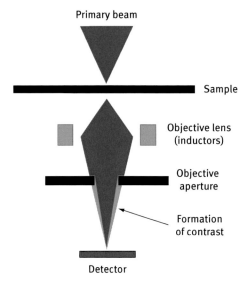

Primary beam

Sample

Objective lens (inductors)

Objective aperture

Formation of contrast

Detector

Figure 3.27: Beam path of the electrons after passing through the sample inside of a TEM. Stronger scattered electrons are removed with an objective aperture. This gives contrast for different materials and material thickness.

When an electron beam is focused on the sample during TEM measurement, the electrons are scattered by passing through the specimen. The amount of scattered electrons increases with the thickness of the sample, whereas the scattering angle increases with the atomic number of the investigated region (e.g., metal particles in electrospun fibers). Therefore, metal particles appear darker than the matrix, for example. This so-called bright-field mode is possible by using an objective aperture, which removes the stronger scattered electrons out of the pathway (Figure 3.27). The bright-field mode is the most common way for the investigation of a sample but one can also postpone the objective aperture for the exploration of the stronger scattered electrons, whereby the brighter and darker regions are switching. This is called dark-field mode. Crystalline structures also provide inherent contrast (diffraction contrast) and are seen as patterned dark regions in TEM images. As opposed to the formerly mentioned contrasts the diffraction contrast is dependent on the tilt angle of the specimen (*Bragg's law*), which can be adjusted. According to how the crystalline lattice is tilted to the incident beam, different atomic planes satisfy the Bragg's law, whereby the appearance of the crystalline structure is changing and a differentiation between other types of contrasts can be easily made. This means that, at high resolutions, even the crystal structure of particles in electrospun fibers can be determined. Due to constructive and destructive interferences well-ordered crystal lattices appear as bars. Because of the low resolution of conventional TEM the fine structure of crystal lattices is sometimes difficult to see. The observation of these fine crystal-lattice structures is possible by use of a high-resolution TEM (HRTEM) and most often a high acceleration voltage of 200–400 kV. Additional advantages emerge in the succession of the tilt-ability of the specimen by using supplementary software and computational power, whence 3D objects can be calculated, using images of different tilt angles. This will give nice 3D images of electrospun blend or composite fibers.

Sample preparation

Typical specimen holders for TEMs are made of copper grids. The sample deposited on this grid between the intervening spaces is used for the analysis (Figure 3.28). The electrospun fibers can be directly collected on a copper grid. Furthermore, already made fibers can be put between folding grids (a double grid for jamming the specimen). (Caution! The fibers sandwiched between the grids are loose and do not adhere strongly. This can make the focusing procedure difficult.) Since electrospun fibers can have diameters greater than 200 nm, it is necessary to use an acceleration voltage of at least 300 kV to penetrate the sample (Figure 3.29).

To explore the cross section of a fiber, oriented free-standing fibers are needed. This can be done by electrospinning on a fast rotating disk, which is wrapped, for example, with baking paper for an easy removal of the fibers. The fiber cross sections are cut by embedding the oriented fibers in a two-component epoxy adhesive. Generally, few hours are required for the complete curing (cross-linking) of the embedding epoxy matrix and this is followed by cross section cutting by ultramicrotome.

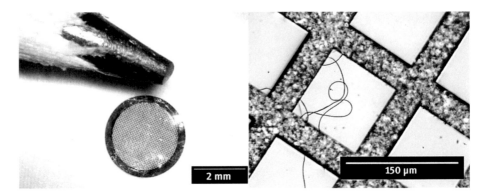

Figure 3.28: The size of a copper grid in comparison to a pencil (*left*) and electrospun fibers on a TEM grid with the carbon support film. The film is not absolutely necessary, but it lowers the movement of the fibers (*right*).

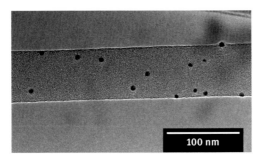

Figure 3.29: TEM image of an electrospun polyimide nanofiber with silver nanoparticles, using an acceleration voltage of 300 kV.

The obtained cross sections should be thinner than 100 nm. The free floating cut cross sections on the surface of a liquid can be fished using a TEM grid. If the sample is too soft, the cuts can even be performed by the utilization of liquid nitrogen (cryomicrotome) so that the temperature is far below the glass transition temperature of the polymer, whereby the stiffness rises.

In cases of polymer blend or copolymer fibers the differences of contrasts might be too small for a distinction in TEM, because the thickness and atomic numbers are almost in the same range. Therefore, different functional groups can be stained exclusively by oxidizing agents with higher atomic numbers. For example, osmium tetroxide (OsO_4) can be used for staining of double bonds (Scheme 3.1).

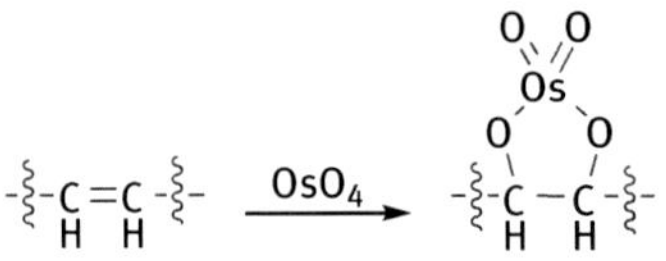

Scheme 3.1: Reaction of osmium tetroxide with double bonds.

The staining with OsO_4 can either be carried out from a solution or by vapor deposition, since it has a small vapor pressure. For example, for staining with OsO_4, the sample deposited on a TEM grid is placed inside a small round-bottom flask. After adding a crystal of OsO_4 the flask is closed overnight. The staining can be fastened (in minutes) by using a diaphragm pump.

Another most common oxidizing agent used for staining purposes is Ruthenium tetroxide (RuO_4), which is stronger and therefore less selective (Scheme 3.2). It stains almost all kinds of functional groups that can be oxidized (e.g., aromatics and double bonds).

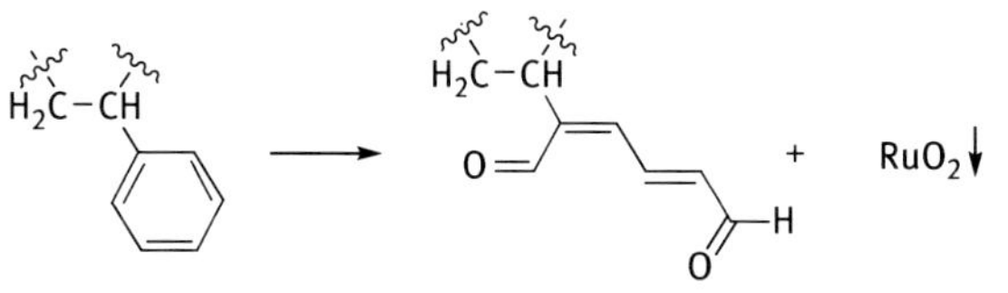

Scheme 3.2: Reaction of ruthenium tetroxide with aromatics.

The staining process with RuO_4 can be performed as with OsO_4 from the gaseous phase. However, RuO_4 is produced in situ, by combining a solution of sodium hypochlorite (NaClO) with ruthenium(III) chloride ($RuCl_3 \cdot xH_2O$) (Scheme 3.3). For staining with RuO_4, combine 0.5 ml of a 10 wt% solution of NaClO in water with 50 mg of $RuCl_3 \cdot 3H_2O$ in a small culture dish. Place the specimen (located on a TEM grid) beside the culture dish and cover everything for 15 min with a petri dish.

$$RuCl_3 \cdot 3\,H_2O \; + \; 4\,NaClO \; \longrightarrow \; RuO_4 \uparrow \; + \; 4\,NaCl \; + \; 3\,Cl^- \; + \; 3\,H_2O$$

Scheme 3.3: Reaction of sodium hypochlorite with ruthenium(III) chloride to produce ruthenium tetroxide.

3.2.2 Molecular structure

3.2.2.1 Infrared spectroscopy

Infrared spectroscopy is a good and fast way for structural characterization providing information about functional groups and molecular orientation in fibers [25]. The great advantage of this technique lies in the manifold and easy opportunities for the sample preparation. Through methods such as attenuated total reflection (ATR), the specimens can be examined as prepared in liquid and solid state. Therefore, spinning solutions and also fiber mats can be analyzed and the differences between those can be easily studied. Nevertheless, there are even ways to measure gaseous phases as well, but this is not necessary in cases of electrospinning.

The functional principle of the IR spectroscopy is based on the absorbance by functional groups in the IR range (usually in the range of 4,000–400 cm^{-1}). This indicates the presence or the absence of double bonds, amines, aromatics, nitriles, organic acids, alcohols, hydrogen bonds, etc., and can be used for a qualitative and quantitative analysis, respectively.

The light can be defined by the following values (Table 3.2):

Table 3.2: Different values to describe the energy of light.

Value	Meaning	Unit	Formula
Frequency	Quantity of oscillations within a defined time scale	(s^{-1}) or (Hz)	$f = \frac{c}{\lambda}$
Wavelength	Distance for one oscillation	(m) (SI unit) or (cm)	$\lambda = \frac{c}{f}$
Wavenumber	Quantity of oscillations within a defined length scale	(m^{-1}) (SI unit) or (cm^{-1})	$\tilde{v} = \frac{1}{\lambda}$ $\tilde{v} = \frac{f}{c}$

Now-a-days the general energy specification for IR spectroscopy is done in cm^{-1} because the value increases linearly with the energy, which is given by the following equation:

$$E = h \cdot f = h \cdot c \cdot \tilde{v} \tag{3.25}$$

E energy (J)
h Planck's constant (6.626 × 10^{-34} J s)
f frequency (s^{-1})
c speed of light in vacuum (2.998 × 10^{10} cm/s)

In the IR region, the electromagnetic waves couple with the molecular vibrations. If the energy of the incident electromagnetic radiation is absorbed (e.g., in fiber mats), the location or the bond lengths between atoms in a molecule can vary. According to that, a certain proportion of the absorbed energy can be transformed into kinetic energy. This is only possible if a dipole moment changes or is induced. That means, if a vibration is symmetrical to a center of symmetry, the dipole moment will not change and no light is absorbed. Simultaneously, the polarizability of the molecule rises in such situations, which results in a higher scattering of the incident beam and can be measured by complementary techniques, for example, Raman spectroscopy. This phenomenon is called the "rule of mutual exclusion" and implies that vibrations in molecules with a center of symmetry cannot be IR active but at the same time Raman active (Figure 3.30).

There are two different types of IR spectrometer for the laboratory analysis (Figure 3.31): Dispersive infrared spectroscopy and Fourier transform infrared spectroscopy (FT-IR, often shown with a Michelson interferometer).

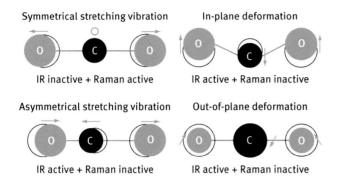

Figure 3.30: Schematic illustration of the different vibrational modes of CO_2.

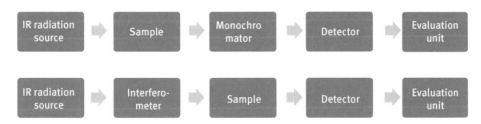

Figure 3.31: Beam path of the dispersive infrared spectroscopy (*above*) and of the Fourier transform infrared spectroscopy (*below*). The difference is the location of the spectral devices (monochromator and interferometer) relating to the sample position. A monochromator can be placed in front or behind the sample.

Since the nature of the instrument and the influence of the environment have to be excluded, it is always necessary to measure the background. Devices for the dispersive infrared spectroscopy have mostly a two-section beam path, one for the sample and one for the reference, whereas the FT-IR spectrometers are typically single-sectioned. The latter is the more modern way of IR spectroscopy, because it has some advantages such as high signal-to-noise ratio and fast acquisition.

In dispersive infrared spectroscopy the different wavenumbers of the incident beam are unbundled by a swiveling diffraction grating (or a prism), whereupon only a certain wavelength can pass a slit and the intensity is recorded by a detector. To shift between the reference and the sample beam, an optical chopper is used to focus only one irradiation onto the detector (Figure 3.32).

FT-IR spectrometers are using interferometers (such as Michelson interferometers) to generate constructive and destructive interferences from two signals. These interferences yield to a function of intensity and an optical path difference that is called interferogram. To change the optical path difference into the wavenumber, a Fourier transformation is employed (Figure 3.33).

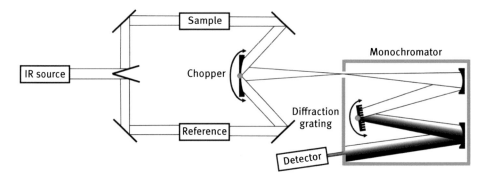

Figure 3.32: Beam path in dispersive IR spectroscopy.

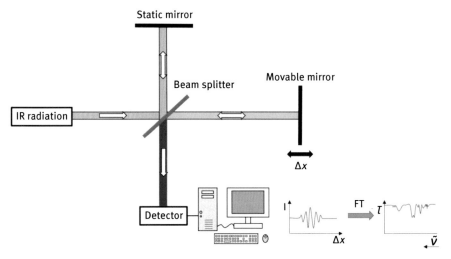

Figure 3.33: Working principle of a Michelson interferometer.

The measurement of (IR-) transparent or (IR-) reflecting samples never posed a problem by applying transmission or reflection modes, but other solid substances (like typical electrospun fibers) had to be pressed within a KBr pellet/NaCl pellet, or they had to be dissolved in an appropriate solvent, respectively. In contrast, the attenuated total reflectance (ATR) technique is an easier way for examining fibers, as it operates with a single crystal that is a reflective element (high refractive index). Inside of the crystal, evanescent waves are formed, which decay exponentially at the surface, but there is still a certain amount of energy that can escape the surface (as according to the particle in a box model) (Figure 3.34). If a sample directly touches the surface, some wavelength can be absorbed (specific wavelength of the IR-active functional groups) and leads to the decrease in the intensity of these wavelengths. This technique enables a direct measurement of solid and liquid samples, such as fiber mats or spinning solutions. Since electrospun fibers are in nanometer range, multilayer fiber mats are even measurable in transmission mode, but empirically strong baseline corrections need to be done, which can sophisticate the result. As-received fiber mats with a flexible support (e.g., aluminum foil, backing paper or carrier fleece) can be utilized with the ATR approach, since the penetration depth is only a few micrometers, so the material of the fiber support will not be measured.

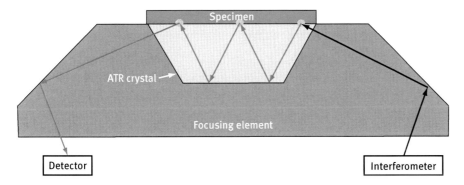

Figure 3.34: Pattern of the ATR method. The incident beam is reflected inside of the ATR crystal and can interact with the specimen at the surface.

Quantitative IR analysis

For a quantitative determination of a substance concentration in electrospinning solutions or in fiber mats, the absorbance (or extinction, not absorption) is used for a signal of a functional group, which is a component of the molecule to be investigated. Therefore, the Lambert–Beer law is used (Equation 3.26):

$$E_\lambda = -\log_{10}(T) = -\log_{10}(1-A) = \varepsilon_\lambda \cdot c \cdot d \tag{3.26}$$

E_λ absorbance or extinction of the substance for a specific wavelength (arbitrary unit)
T transmittance (%)
A absorption (%)
ε_λ molar absorptivity or molar extinction coefficient at a specific wavelength (L/mol m)
c concentration of the substance (mol/L)
d thickness of the sample (m)

The transmission can be written as:

$$T = \frac{I}{I_0} \tag{3.27}$$

I intensity of the transmitted beam (W/m²)
I_0 intensity of the incident beam (W/m²)

The extinction is directly proportional to the concentration of the substance and is the negative logarithm to the base 10 of the transmission. There are mainly two different ways for the determination of the transmission: the difference in intensity or the area of the signal (Figure 3.35).

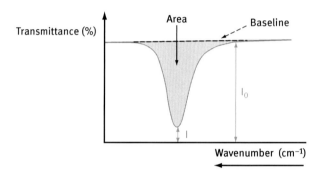

Figure 3.35: Scheme of the graphical determination of the transmittance, which can be performed either by the area under the signal (integral) or by the intensity difference.

With known ratios of concentration, a calibration line needs to be compiled, which contains at least five known concentrations for a good reliability (Figure 3.36).

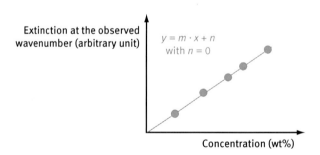

Figure 3.36: Scheme of a graph for a quantitative IR analysis.

In order to reduce the error, the target unknown concentration should be within the calibrating values. Another error could arise from different observed thickness of the sample, so the spectra should be normalized for the intensity of additional groups (e.g., of the matrix). The limit of quantitation for the instrument is measurable by multiple tests of the same concentration. If the error is too high (e.g., $\geq 10\%$), the limit of concentration is reached. For the instance of concentration determinations in electrospun fibers, the calibration can be accomplished with the calibration of polymer films or blends. Those specimens can be prepared directly with the spinning solution.

When the fibers are collected on a substrate such as aluminum foil and backing paper, they can be directly used for FT-IR–ATR measurements.

If the thickness of the fiber mat is big enough, it can be peeled off from the support and placed directly on the sample holder. In this case the transmission mode of the spectrometer is used. As previously mentioned, this method empirically requires more intense baseline corrections and impacts of the sample thickness strongly influence the result.

A general chart of the expecting signals is shown in Figure 3.37. A detailed summary of the expected signals can be found in the literature [25].

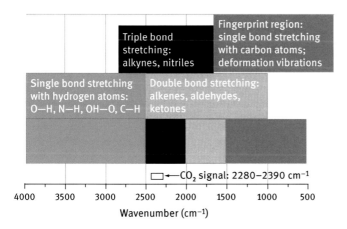

Figure 3.37: Rough estimation of IR signals.

3.2.2.2 Thermogravimetric analysis

Thermogravimetric analysis (TGA) is used to correlate the mass loss or gain on heating a sample with a constant heating rate, with physical and chemical changes occurring. The mass loss or gain can also be observed at different time intervals at a constant temperature. Physical transitions such as vaporization, evaporation, sublimation, and chemical reactions like decomposition, reduction (e.g., of metal oxides), combustion (if oxygen is available) can be monitored using TGA.

The method is used for electrospun fibers mostly but not limited to the determination of the thermal stability, solvent residues, and weight percentage of the different components in case of composite or blend fibers. Typically, the combustion chamber is flooded with an inert gas such as nitrogen or argon and weight losses as a result of physical or chemical phenomenon are measured, but it is also possible to use air or oxygen to examine the degradation or oxidation behavior under defined conditions. With the latter the weight increases, since the oxygen is integrated into the molecular structure of the sample. A scheme for a TGA setup can be seen in Figure 3.38.

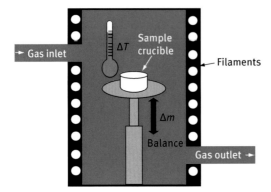

Figure 3.38: Illustration of a TGA.

To gain more information about the decomposition products, it is often possible to collect the exhaust fumes and analyze them further with a gas chromatograph–mass spectrometer (GC–MS) or with an infrared spectrometer (IR).

For an evaluation of the obtained measurement curve, the change in mass for the distinct decomposition stages is calculated, as well as the remaining masses, called char yield. The calculation of the different decomposition temperatures can be realized in two different ways: by determining the temperature, at which a specified percentage of the degradation took place [e.g., the temperature at which a 5% change in mass of the initial quantity $(T_5\%)$], extrapolated initial (T_i) and final (T_f) decomposition temperatures or by using the first derivative at which the maxima correspond to the temperature at which the rate of decomposition is at its maximum (T_{max}) (Figure 3.39).

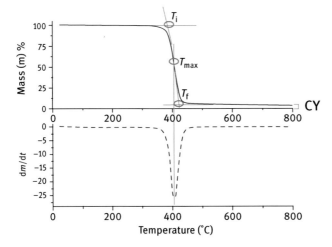

Figure 3.39: Thermogravimetric (TG) and differential thermogravimetric (DTG) curves depicting single-step degradation. Thermal stability is assessed by noting extrapolated initial and final decomposition temperatures (T_i and T_p), temperature at which the rate of decomposition is maximum (Tmax) and char yield (CY).

Depending on the polymer used for spinning, more than one degradation step can be seen in TGA. A hypothetical multistep thermal degradation behavior is shown in Figure 3.40. Both curves show three steps degradation but the steps are distinct in curve A and overlapping in curve B. The overlapping steps can be seen if the heating rate or the amount of the sample used for analysis is too high. The TGA analysis is influenced by several factors such as heating rate, sample amount and form, and gas flow rate. Therefore, it is necessary to provide this information while reporting results. For comparative analysis among a series of samples it is recommended to keep all the parameters same.

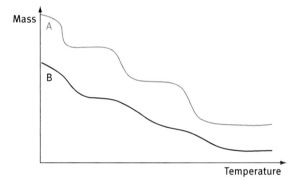

Figure 3.40: Hypothetical TG curves showing three-step degradation behavior. The steps are clearly separated in curve A, whereas curve B shows overlapping steps. Slowing down the heating rate and using less amount of the sample (5–10 mg) might resolve the steps separately.

Sample preparation

The fibers should be dried thoroughly before measurement. The first choice concerns the right material for the crucible. For high temperatures, a crucible made of corundum (Al_2O_3) fits best, but substances such as salts, oxides, or minerals in composite fibers, could react with corundum at elevated temperatures, so an aluminum pan needs to be taken, which is only practicable for temperatures below 650 °C (melting temperature of aluminum: 660 °C) (Figure 3.41). For weighing the sample in a crucible, a balance with a minimum precision of 0.1 mg should be used.

For TGA measurement, cut out a small piece of the fiber mat and put it inside of the crucible with the help of tweezers. The weight should be around 5–15 mg, but do not overfill the crucible. Note the weight of the sample. To draw comparisons between different samples, the amount of samples used for measurement should be similar. By using a plugger, compress the sample to the bottom of the crucible to minimize thermal gradients. Place the specimen inside of the TGA. Choose a matching heating profile. A temperature range of 25–650 °C fits an aluminum pan and 25–800 °C matches a corundum crucible. The heating rate is typically 10 K/min or 20 K/min.

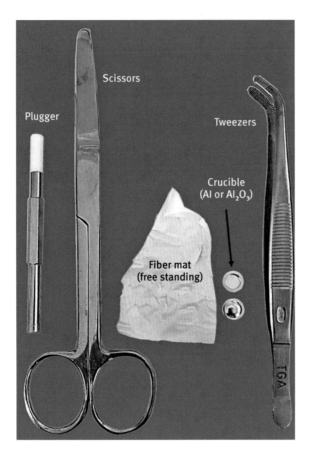

Figure 3.41: Required tools for a TGA measurement.

3.2.2.3 Differential scanning calorimetry

A differential scanning calorimetry (DSC) is a powerful and readily available simple instrument for the detection of thermal transitions in materials including polymers. It enables the measuring of quantities such as the glass transition temperature (T_g), (re)crystallization temperatures, melting temperatures (T_m), and mesomorphic transition temperatures (in liquid crystals) and delivers information, if a blend is miscible or not. Since the T_g is dependent on the molecular weight and purity of the polymer, further investigations are possible (Figure 3.42). The T_g dependency on the molecular weight, the degree of branching, and tacticity (refer to Chapter 5) is the reason, why ranges for the T_g and T_m are often indicated for a specific polymer.

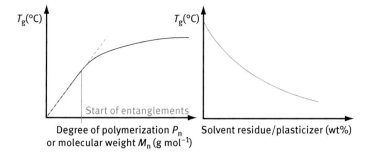

Figure 3.42: Dependence of glass transition temperature (T_g) on molecular weight and solvent residue: T_g increases with the molecular weight (or degree of polymerization) till a plateau is reached and decreases with increased residues of a solvent.

The operation principle of a DSC is the simultaneous heating of a sample and a reference substance and comparing the heat flow between them. The reference substance should not undergo any thermal transitions (e.g., decomposition) in the temperature range in which the measurement is carried out. Generally, empty aluminum crucibles that are otherwise used for holding samples for DSC measurements are used as reference materials. If there is a difference in the temperature between the sample and the reference due to phase transitions in the sample during measurement, an additional temperature control unit caters for the compensation. The energy for this compensation is proportional to the difference between the heat flow of the sample and the reference (Equations 3.28 and 3.29).

$$\Phi_t = \Phi_s - \Phi_r \propto \Delta T \tag{3.28}$$

Φ_t total heat flow (W) or (J/s)
Φ_s heat flow of the sample (W) or (J/s)
Φ_r heat flow of the reference (W) or (J/s)
ΔT difference in temperature between sample and reference (K)

Together with the mass of the sample, the heating rate and the temperature of the sample, a typical DSC curve can be obtained (Figure 3.43).

$$\Phi_t = \frac{\delta Q}{dt} = C\frac{dT_s}{dt} \tag{3.29}$$

δQ quantity of heat (J)
dt difference in time (s)
dT_s difference in temperature of the sample (K)
C specific heat capacity (J/K kg)

In order to emphasize the continuity of the pressure in the process, the heat capacity at constant pressure is used:

$$C_p = \frac{dH}{dT} \qquad (3.30)$$

dH difference in enthalpy; equivalent to a feed of heat at constant pressure: δQ_p

According to Equation (3.29) the integration of a signal over the temperature results in the change of enthalpy for the specific transition.

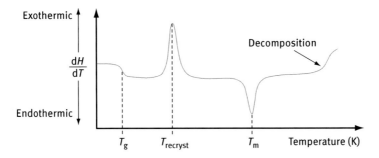

Figure 3.43: Common DSC heating profile of semicrystalline polymers. There is no melting temperature and recrystallization temperature in cases of amorphous polymers.

Due to the process of nucleation and growth, a melt of a semicrystalline polymer typically shows supercooling, so the crystallization temperature (T_{cryst}) is below the melting temperature (T_m) (not to be confused with the recrystallization temperature $T_{recryst}$). This phenomenon is connected to the cooling rate. Between T_g and T_m of semicrystalline polymers, a recrystallization can appear. For amorphous polymers, the DSC curve shows only a glass transition (T_g) as a shift in the base line. The glass transition temperature is commonly defined as the temperature at $0.5 \times \Delta C_p$ for the specific transition, but frequently the onset temperature is also used, so one has to check the details of the data in hand to bypass errors.

From the integral of the melting signal, the enthalpy of fusion can be obtained. By dividing a theoretical value for the same polymer with a 100% crystallinity (can be found in the literature) with the measured quantity, the crystallinity of the analyzed polymer can be evaluated (Equation 3.31). This can be interesting for determining the change of crystallinity due to electrospinning processes. Other ways to identify the crystallinity of a polymer are wide-angle X-ray scattering (WAXS), density measurements and sometimes methods of infrared spectroscopy (IR spectroscopy).

$$w_c = \frac{\Delta H_p \cdot 100\%}{\Delta H_{cryst}} \tag{3.31}$$

w_c crystallinity of the present polymer (%)
ΔH_p enthalpy of fusion for the present polymer
ΔH_{cryst} theoretical value for the polymer with a 100% crystallinity

Another advantage of DSC is the assessment, if a blend is miscible or not. Miscible blends have only one T_g, whereas immiscible blends show more than one glass transition temperature. This is extremely important, if blends are used for electrospinning, in order to modify mechanical properties of fibers or just to improve the spinnability. In immiscible blends, microdomains are built, which manifest in completely different physical properties (thermodynamical, mechanical, optical properties) as to miscible blends. So fibers with the same components can show different physical and mechanical properties.

Sample preparation

The fibers should be dried thoroughly before the measurements and should not be touched without gloves (the same applies to the specimen holder: commonly an aluminum DSC pan). The tools for DSC measurement are shown in Figure 3.44. The best way is to perform a TGA measurement in front of a DSC test, so that the correct temperature range for DSC measurement can be selected, since temperatures during DSC measurements should not exceed the decomposition temperature of the fibers. In general, the thermal transitions in DSC are analyzed in the second heating–cooling curves. The first heating cycle is used to erase the thermal history of the samples.

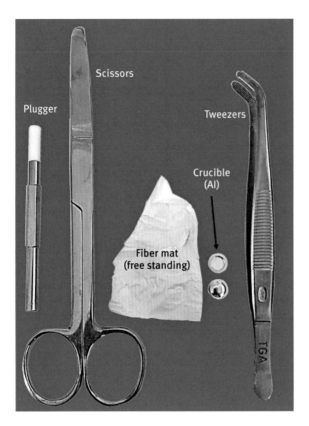

Figure 3.44: Required tools for specimen preparation in DSC measurements.

A balance with a minimum precision of 0.1 mg should be used for weighing the fiber mats. Cut out a small piece of the fiber mat and put it inside of the lower half of a DSC pan with the help of tweezers. The weight should be around 5–15 mg, but do not overfill the pan. To draw comparisons between different samples, the amount of weight should be similar. Note the weight of the sample. Compress the sample to the bottom of the DSC pan to minimize thermal gradients. Cover the filled side of the DSC pan with the lid and connect the two parts via cold welding (typically, a pan sealing press is used). Perforate the lid, so potential gases can escape and will not burst the specimen holder. Place the specimen inside of the DSC. Choose a matching heating and cooling profile. A heating and cooling rate of 10 K/min is good for a first trial.

3.2.2.4 Tensile testing

For a mechanical characterization of electrospun fibers, it is possible to perform tensile tests. The most important results of these tests are the Young's modulus, the ultimate tensile strength, and the elongation at break and toughness (Table 3.3).

Those quantities serve for an adequate assessment for the mechanical properties and can be measured on fiber mats and even on single nanofibers. Nevertheless, extremely sensible load cells and special treatments are needed for the analysis of single nanofibers. An illustration of the important quantities determined during tensile testing can be found in Figure 3.45. There are of course additional quantities to be named such as yield strength and offset yield strength. The need to measure all of them depends on the application.

Table 3.3: Important quantities in tensile tests.

Symbol	SI unit	Character	Meaning
E	Pa	Young's modulus; tensile modulus; elastic modulus	Quantity of elasticity; resistance against elastic deformations
σ_M	Pa	Ultimate tensile strength; tensile strength; ultimate strength	The maximum stress a material can endure
$\varepsilon_M; \varepsilon_B$	%	Maximum elongation; elongation at break	The maximum elongation a material can withstand before breaking
T	Pa	Toughness	Total energy that a material can absorb

The Young's modulus is calculated from the linear elastic region, which is the first slope of a tensile test. Here the *Hooke's law* can be applied:

$$\sigma = E \cdot \varepsilon \tag{3.32}$$

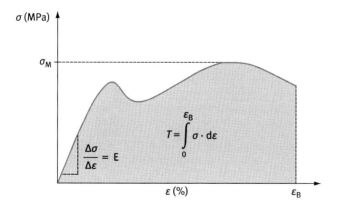

Figure 3.45: Schematic representation of a stress–strain curve and the most important corresponding quantities. The behavior of brittle materials often shows only the linear elastic region, whereas elastomers have higher strain hardenings and a pronounced elongation.

For the calculation of the toughness, the area under the stress–strain curve is used. A significant circumstance of highly elastic materials – such as rubbers – is the pronounced toughness, since the elongations at break are augmented as well. On the contrary, brittle matters (the same applies to brittle fibers) have high elastic moduli but low energy absorptions. The stress–strain curve for brittle fibers will end after the first slope and elongation at break is very low.

For the test, a specimen (e.g., fiber mat) with a defined cross section (there are different standards available and testing is carried out according to these standards) is clamped and stretched. Thereby the emerging force is determined and calculated into a stress quantity (Equation 3.33):

$$\sigma_n = \frac{F}{A} \tag{3.33}$$

F emerging force (N)
A initial cross section of the specimen (mm²)
σ_n nominal stress or engineering stress (N/mm²) or (MPa)

In the process, the elongation of the sample is calculated according to:

$$\varepsilon = \frac{\Delta L}{L_0} \cdot 100\% \tag{3.34}$$

ε elongation (%)
ΔL change in length
L_0 original length

During the test, the diameter of the sample decreases when it is stretched. That is why the stress is called nominal stress or engineering stress in this context. If the size of the sample is continuously evaluated, a true stress can be used, which is higher than the nominal stress since the diameter of the sample decreases during the tests. An illustration of the pathway by observing the real diameter can be seen in Figure 3.46. There the ascents of the curve before failure ought to be noticed instead of a drop as in Figure 3.45.

After the yield point, which can be seen as the border between elastic and purely plastic deformations (totally irreversible deformations), a drop in the curve can (sometimes) be recognized. This is due to the start of necking. Before the yield point, the whole clamping length (the length of the sample between the clamps) participates uniformly in the elongation. Subsequently the elongation is restricted to the regions of necking. For that reason, the mechanical properties are not independent from the shape of the specimen and from the clamping length. With an increasing length-to-diameter ratio (e.g., increasing clamping length with constant diameters of the sample), the percentage of the necking region decreases, so the fraction of elongation for the plastic region decreases as well. On the contrary the elongation for the elastic region increases, as the stretching is uniformly distributed over the sample, as already mentioned. Especially for elastomers, a necking cannot be observed, since the polymer chains are networked and free motions are impossible. In this case, the elongation at break is constantly enhanced with the clamping length.

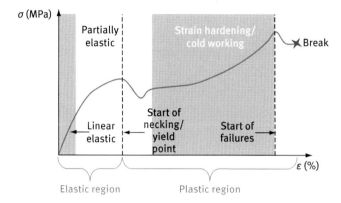

Figure 3.46: Scheme of a true stress–strain curve with annotations about the molecular proceedings (different stretching regions). A curve with all regions is shown here. In cases of brittle materials, only the first slope will be recognized, whereas elastomers show higher elongations and hence the region of the strain hardening is wider. The start of the plastic region (totally irreversible deformation) is called "yield point" and is not always visible.

Following the necking, another rise of the stress can be observed. For the reason of polymer chain stretching, the alignment of the molecules has to be enhanced equally. Orientations in the direction of the load are always connected to higher tensile strengths, whereas strength in the orthogonal direction decreases. That is even the basis, why a nonwoven fiber mat with randomly oriented fibers are weaker than a fiber mat with aligned fibers (in the direction of fiber alignment), where the total load is equally distributed over the entire number of fibers.

Further features of fibers are higher ultimate tensile strengths and pronounced elastic moduli as opposed to bulk materials, in general. However, theoretical data are always higher than empirical quantities, since there are no real materials that are completely defect free [26].

Continuing with the mechanical dependence on the morphology of fibers, lower diameters show nonlinear increments in the Young's moduli and ultimate tensile strengths (Figure 3.47) as also described in Chapter 2. This was first observed by A. A. Griffith [27], who recognized this effect in tests with glass fibers. The same effect applies to electrospun fibers, where different explanations are discussed, such as decreasing probabilities for defects, higher strain rates (which disembogue in higher molecular orientations) and higher impacts of the surface morphologies [28].

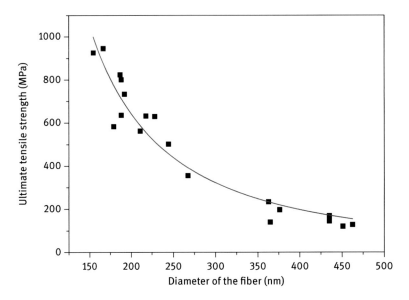

Figure 3.47: Single fiber tests on electrospun poly(vinyl alcohol) (PVOH) fibers. With decreasing diameters, the ultimate tensile strength increases nonlinearly. For the evaluation of the diameter, every fiber needs to be examined by SEM.

These investigations have to be carried out on single fibers, which is possible with extremely sensible tensile testing machines (nano tensile tester; load measuring range from 0.01-98.07 mN) or atomic force microscopy (AFM) (three-point bending tests and nanoindentations). For mechanical testing using a nano tensile tester the single fibers are collected on a metal frame and transferred carefully to a cardboard/ paper frame for mounting on a tensile tester as shown in Figure 3.48. A double-sided electrically conductive tape is utilized for fixing the single fiber on a cardboard frame as the fiber piece left on the frame after testing can be used for diameter analysis by scanning electron microscopy (SEM). The diameter in single fiber tests needs to be examined for every fiber since even variations of few nanometers in the measurement of the fiber diameter can lead to large errors of several percentages. In the process, the geometry of the fiber, as another critical factor, can be checked. This is important for the underlying calculations of the cross section area and for the mechanical characteristics by itself, because the necking and failure behavior is different for every shape (round, ribbon-like, beaded, pored, etc.). After mounting the single fiber attached to the cardboard/paper frame, the sides of the frames are carefully cut before starting the measurement.

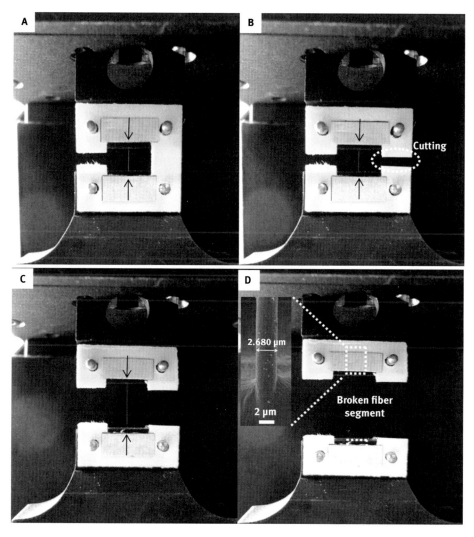

Figure 3.48: (A) Mounting of paper frame with a single fiber on nano tensile testing machine; (B) cutting the paper frame from the side and starting tensile testing; (C) stretched fiber during testing; (D) the fiber segments left on conductive tape after tensile testing are used for SEM measurements for diameter determination. [Reprinted with permission from ACS: *Appl. Mater. Interfaces* **2014**, *6*, 5918. Copyright ACS (2014).]

In a three-point bend test method for the determination of elastic modulus the single fiber is fixed over a groove on a substrate such as silicon wafer from both ends and a force is applied via AFM at a constant loading rate at the midspan of fiber (Figure 3.49).

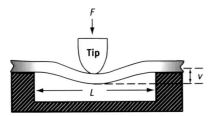

Figure 3.49: Schematic of a three-point bend test for the determination of elastic modulus of a single fiber using AFM. The midspan of a fiber of suspended length L is deflected by an AFM tip. [Adapted with permission from *Appl. Phys. Lett.* **2004**, *84*, 1603. Copyright American Institute of Physics (2004).]

The fiber is considered as an elastic beam and the elastic modulus is determined by observing the deflection of the beam at midspan using the following equation:

$$E = \frac{FL^3}{192vI} \tag{3.35}$$

where F is the maximum force load applied, L is the suspended length of the fiber, v is the deflection at the midspan and $I = \pi D^4/64$ (D = diameter of fiber). While carrying out three-point bending tests one should be aware of the assumptions made. The fiber was assumed as an elastic beam with uniform diameter all throughout the length and the load applied did not cause any indentation.

Another method of measuring elastic modulus of single fibers is by nanoindentation using AFM. A nanometer-sized AFM cantilever tip is indented into a fiber fixed on a hard and flat surface. The applied force is calculated from the nominal spring constant of cantilever. The indentation depth (δ) should be well below the thickness of the sample in order to avoid errors coming from the substrate surface. Different models are used for the calculation of elastic modulus assuming either the fiber surface flat (valid only when the cantilever tip is much smaller than the fiber radius) or fiber and cantilever tip as spherical bodies in elastic contact [28–31]. The obtained modulus values should be corrected for surface roughness. Generally, the electrospun fibers are not completely smooth. The surface roughness can lead to decreased contact pressure for a particular load as the contact area will increase. The indentation method might also give incorrect values for very thin samples (< 200 nm) due to the effect from the substrate. The AFM tip might deform the substrate in such a case.

3.3 Interesting to know

– There are many polymers that undergo a sharp change in property such as color, shape and size, conformation and hydrophilicity–hydrophobicity on change in external stimuli. The external stimuli triggering a change in the polymer properties can be temperature, pH, light, ionic strength, magnetic and electric field and many others. Such polymers are called "smart," "intelligent" or "responsive" polymers. When we talk about the solubility of polymers in solvents (either organic or water), it is temperature dependent. The most common notion among students is to increase the solubility of a polymer by heating. You might get surprising results if you try to dissolve some polymers by heating, especially in water, although it is also possible in organic solvents. Instead of seeing a clear solution you might end up in seeing a precipitation on increase in temperature. The polymers that change their hydrophilicity–hydrophobicity/ solubility in water or organic solvents with temperature belong to the smart polymers titled as thermoresponsive polymers. Such polymers are of two types: the one that phase-separates (precipitates) on heating above a critical temperature [lower critical solution temperature (LCST)-type polymers] and the other one that phase-separates on cooling [upper critical solution temperature (UCST-type)]. PNIPAm is one of the polymers that will precipitate out on heating above 32 °C. On the other hand, some other polymers like copolymers of acrylamide (Am) and acrylonitrile (AN) will not dissolve so easily in water and require heatings above their critical temperatures, depending on the ratio of Am and AN. They are examples of polymers with UCST-type behavior. The critical temperature at which phase separation takes place depends on many factors, such as polymer type (chemical structure), molar mass, molar mass dispersity, chain ends and concentrations. Experimentally, temperature-dependent phase separation of polymers in water or organic solvents can be tested by turbidity photometer/UV–Vis spectrophotometer, micro-DSC, temperature-dependent NMR and viscosity measurements, among other methods.

Similar to temperature-dependent solubility, you might observe that the dissolution of some polymers is also dependent on pH. Do not be surprised if some polymers dissolve at a particular pH but precipitate out on slight change in pH. Such polymers are also called pH-responsive polymers and most common examples are polymers that have functional groups such as acids (–COOH) and amines (–NH$_2$). Details about responsive polymers can be read in review articles [5, 32].

– Sometimes for the identification or characterization of special fiber morphologies, for instance a side-by-side structure, it is beneficial to use other supporting techniques such as confocal microscopes (Figure 3.50) or confocal Raman microscopes in addition to scanning electron microscopes. In that case the polymer solutions are generally dyed with specific fluorescent dyes before spinning. An illumination of the dyed sample with light of a particular wavelength, leads to the appearance of the fluorescent color. Raman analysis can help in confirming the different chemical structures on the two sides of the fibers. Raman confocal microscopy is a coupling technique that combines a Raman spectrometer to a confocal microscope. It is a very nice technique for the visualization of samples by fluorescent colors and simultaneous determination of the chemical structure by Raman analysis, without elaboration and any extra efforts in sample preparation [33].

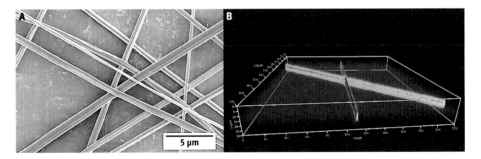

Figure 3.50: Side-by-side [silk fibroin-poly(L-lactide)] fibers as seen by (A) SEM, (B) confocal microscope [Uranine and Rhodamine B was added to poly(L-lactide) and silk fibroin spinning solutions, respectively].

– Many times, conventional DSC is not helpful in making comments about phase transitions due to overlapping peaks, solvent evaporations, and weak transitions, etc. In such cases, a modified DSC – modulated DSC (MDSC) – could be helpful. MDSC uses the same setup, but different heating and cooling profiles. In conventional DSC, a linear heating/cooling rate is applied, but in MDSC a sinusoidal modulation, characterized by amplitude (A) and angular frequency (ω), is overlaid on the linear heating program, which is given by the following equation.

$$T(t) = T_0 + \beta_0 \cdot t + A_t \cdot \sin(\omega t) \tag{3.36}$$

T_0 initial temperature (K)
β_0 average heating rate (K/s)
t time (s)

The right heating program that gives the best separation and resolution of peaks can be generated by varying parameters like the amplitude. The total heat flow is separated into two parts in MDSC and is described by the equation 3.37. It has two components: heating rate dependent– and absolute temperature–dependent components.

$$\frac{dQ}{dt} = -\frac{dT}{dt}[C_p + f''(t, T)] + f(t, T) \tag{3.37}$$

$\frac{dQ}{dt}$ heat flow (W/s)
$\frac{dT}{dt}$ heating rate (K/s)
C_p heat capacity (J/K)
T temperature (K)
t time (s)
f''(t,T)   heat component – thermodynamic
f(t,T)   heat component – kinetic

Crystallization, cross-linking, decomposition, solvent evaporation, and sublimation are absolute temperature-dependent transitions (nonreversing) and can be easily separated from transitions, like melting and glass transition, which are heating rate dependent and can be reversed by heating–cooling cycles with modulated DSC.

Further relevance for the use of MDSC, especially for electrospun fibers, could be the separation of phenomena like a glass transition from relaxation occurrences. The processing of polymers generally results in internal stresses that are also called thermal history effects. Heating the sample with thermal history beyond the glass transition temperature provides a small endothermic peak in DSC due to the release of internal stresses. Since the endothermic stress-relaxation peak is very close to the glass transition temperature, it can make DSC analysis difficult and can also be misinterpreted for a melting peak.

This is the reason that the results in conventional DSC analysis, of any polymeric material, are taken from the second heating cycle, i.e., the sample is heated in the first heating cycle till a temperature above the glass transition temperature, cooled and reheated in the second cycle, until desirable temperatures with a constant linear heating rate. For many purposes, it is required to study the transitions of as-spun fibers and therefore the DSC curve from the first heating cycle should be interpreted. To avoid misinterpretations due to the reasons highlighted above, it is recommended to use MDSC in case of doubts [34].

– The molar mass of the polymers is expressed as an average value. This is in contrast to the small molecules that are associated with one unique molar mass. For example, the molecules with molar masses 58.08 g/mol, 60.05 g/mol and 64.07 g/mol are acetone, acetic acid and sulfur dioxide. In contrast, the polymers have no unique molar masses. This means a particular molar mass, for example, 10,000 g/mol can be associated with any polymer. Moreover, they are characterized by a distribution of molar masses that depends on the polymerization technique used for the synthesis. During polymer formation, macromolecular chains of different lengths are made and the molar mass is expressed as an average of molar mass of all macromolecular chains. The polymer with macromolecular chains of different lengths is called polydisperse or nonuniform. The most common ways of describing the average molar masses are: number average molar mass (M_n), weight average molar mass (M_w) and viscosity average molar mass (M_v). The ratio of weight average molar mass (M_w) and number average molar mass (M_n) is called a molar mass dispersity ($Đ$) which is more than 1.0 for polydisperse polymers. The polymers with all macromolecular chains of the same length are called monodisperse or uniform polymers. The molar mass dispersity ($Đ$) is 1.0 for such polymers. Synthetic polymers made by anionic polymerization are generally monodisperse as the formation of all macromolecular chains during polymerization starts at the same time due to the rate of initiation that is much faster than the rate of propagation and due to absence of termination and transfer reactions. There are different methods of determining the molar mass of polymers. One of the simplest methods is to determine the intrinsic viscosity (described in Section 3.1) which is correlated to polymer molar mass (M) by Mark–Houwink equation (3.38).

$$[\eta] = k[M_v]^a \tag{3.38}$$

$[\eta]$ intrinsic viscosity
M_v viscosity average molar mass

Other methods of polymer molar mass determination are gel permeation chromatography (also called size exclusion chromatography), light scattering, colligative properties determination, MALDI-TOF (matrix-assisted laser desorption ionization-time of flight), and NMR (nuclear magnetic resonance).

References

[1] B. A. Miller-Chou, J. L .Koenig, *Progress in Polymer Science* **2003**, *28*, 1223–1270.
[2] W. J. Cooper, P. D. Krasicky, F. Rodriguez, *J. Appl. Polym. Sci.* **1986**, *31*, 65–73.
[3] A. C.Ouano, J. A. Carothers, *Polym. Eng. Sci.* **1980**, *20*, 160–166.
[4] C. R. Wilke, P. Chang, *AIChE J.* **1955**, *1*, 264–270.
[5] J. Seuring, S. Agarwal, *Macromol. Rapid Commun.* **2012**, *33*, 1898–1920.
[6] J. H. Hildebrand, R. L. Scott, *The solubility of nonelectrolytes*, Reinhold Pub. Corp., New York, USA, **1950**.
[7] C. M. Hansen, *Hansen solubility parameters: A user's handbook*, Taylor & Francis, Boca Raton, FL, **2007**.
[8] J. P. Teas, *J. Paint Techn.* **1968**, *40*, 19–25.
[9] C. J. Luo, E. Stride, M. Edirisinghe, *Macromolecules* **2012**, *45*, 4669–4680.
[10] C. J. Luo, M. Nangrejo, M. Edirisinghe, *Polymer* **2010**, *51*, 1654–1662.
[11] C. Huang, S. Chen, C. Lai, D. H. Reneker, H. Qiu, Y. Ye, H. Hou, *Nanotechnology* **2006**, *17*, 1558–1563.
[12] T. Uyar, F. Besenbacher, *Polymer* **2008**, *49*, 5336–5343.
[13] S. L. Shenoy, W. D. Bates, H. L. Frisch, G. E. Wnek, *Polymer* **2005**, *46*, 3372–3384.
[14] M. G. McKee, G. L. Wilkes, R. H. Colby, T. E. Long, *Macromolecules* **2004**, *37*, 1760–1767.
[15] J. R. Diasa, F. E. Antunes, P. J. Bartolo, *Chem. Eng. Trans. (CEt)* **2013**, *32*, 1015–1020.
[16] P. Peer, M. Stenicka, V. Pavlinek, P. Filip, I. Kuritka, J. Brus, *Polymer Testing* **2014**, *39*, 115–121.
[17] C. Mit-uppatham, M. Nithitanakul, P. Supaphol, *Macromol. Chem. Phys.* **2004**, *205*, 2327–2338.
[18] C. Wang, H.-S. Chien, C.-H. Hsu, Y.-C. Wang, C.-T. Wang, H.-A. Lu, *Macromolecules* **2007**, *40*, 7973–7983.
[19] L. Li, Y.-L. Hsieh, *Polymer* **2005**, *46*, 5133–5139.
[20] A. Varesano, A. Aluigi, C. Vineis, C. Tonin, *J. Polym. Sci. B Polym. Phys.* **2008**, *46*, 1193–1201.
[21] W. P. Cox, E. H. Merz, *J. Polym. Sci.* **1958**, *28*, 619–622.
[22] E. Abbe, *Archiv f. mikrosk. Anatomie* **1873**, *9*, 413–468.
[23] J. Haus, *Optische Mikroskopie. Funktionsweise und Kontrastierverfahren*, Wiley-VCH, Weinheim, **2014**.

[24] P. W. Atkins, J. de Paula, *Atkins' physical chemistry*, Oxford University Press, Oxford, New York, **2006**.

[25] H. Günzler, H.-U. Gremlich, *IR-Spektroskopie. Eine Einführung*, Wiley-VCH, Weinheim, **2003**.

[26] G. W. Ehrenstein, *Faserverbund-Kunststoffe. Werkstoffe, Verarbeitung, Eigenschaften*, Hanser, München, **2006**.

[27] A. A. Griffith, *Philos. Trans. Royal Soc. A Math. Phys. Eng. Sci.* **1921**, *221*, 163–198.

[28] M. Richard-Lacroix, C. Pellerin, *Macromolecules* **2013**, *46*, 9473–9493.

[29] F. Ko, Y. Gogotsi, A. Ali, N. Naguib, H. H. Ye, G. L. Yang, C. Li, and P. Willis, *Adv. Mater.* **2003**, *15*, 1161.

[30] J. G. Park, S. H. Lee, B. Kim, and Y. W. Park, *Appl. Phys. Lett.* **2002**, *81*, 4625–4627.

[31] E. P. S. Tan, C. T. Lim, *Appl. Phys. Lett.* **2005**, *87*, 123106.

[32] F. D. Jochum, P. Theato, *Chem. Soc. Rev.* **2013**, *42*, 7468–7483.

[33] T. Dieing, O. Hollricher, J. Toporski, *Confocal Raman microscopy*, Springer, Heidelberg [Germany], New York, **2011**.

[34] P. S. Gill, S. R. Sauerbrunn, M. Reading, *J. Ther. Anal.* **1993**, *40*, 931–939.

4 Electrospinning experiments

4.1 Introduction about parameters affecting electrospinning

Although the process of electrospinning is used for getting different targeted fiber morphologies, a typical or simple electrospinning experiment is categorized as successful, when continuous thin fibers, with uniform diameters, without droplets and beads are obtained. For getting uniform continuously long fibers, both electrospinning formulations and machine variables are equally important. There are more than 15 parameters that can influence the process of electrospinning in a crucial way. These parameters decide the fiber morphology and diameter and therefore the resulting macroscopical features.

One can divide these parameters into the following three sections: machine variables, spinning solution characteristics and environmental factors. The applied voltage, the type of collector, the distance between the electrodes, the spinning nozzle diameter and type, and the solution flow rate are some of the important machine variables, influencing the spinning process to a large extent. Spinning solution variables are the type of polymer, solvent polarity and vapor pressure, concentration, viscosity, conductivity, surface tension, and additives (e.g., surfactants or additional salts). Humidity and temperature are the most important environmental variables, having high influences on the spinning process. Modern commercial machines have facilities to control the temperature and the humidity precisely.

Viscosity, concentration, and molar mass are important polymer solution characteristics that decide the fiber formation, the diameter and the morphology. Somehow these parameters are related to each other. They are described in detail in Chapter 3. A polymer with a high molar mass (10^4–10^7 g/mol), an appropriate viscosity (20–300,000 cP) and concentrations (1–20 wt%) is generally used for electrospinning. The idea is to have good polymer chain entanglements already in the solution. Otherwise the probabilities of bead or droplet formations during the spinning process are elevated. The polymer concentration that provides fibers is decided by the molar mass, if other spinning parameters are kept constant. Qualitatively, a high-molar-mass polymer makes a viscous solution at "low" concentrations. For example, for very high-molar-mass polymers, i.e., more than 500,000 g/mol, a concentration of <5 wt% would be sufficient for getting fibers by electrospinning in solvents with at least the vapor pressure of water. On the other hand, for oligomers (low-molar-mass polymers; molar mass less than 20,000 g/mol) very high concentrations (20–40 wt% or even more) might be necessary for getting stable jets during the spinning with the same solvent. Refer to Chapter 3 for further details. An important fact to remember is that nonpolymeric, low-molar-mass molecules, like surfactants, amino acids (diphenlyalanine), and tetraphenylporphyrin, which are capable of self-assembly by supramolecular interactions, could also be electrospun by using highly concentrated solutions, as long as no gelation occurs before the spinning process. Macromolecular chain entanglements or intermolecular forces in solutions are necessary for getting fibers by electrospinning.

The voltage for electrospinning depends mostly on the electrode distance and electrode shape. Normally, an electrode distance of 15–20 cm is used, and the applied voltage is around 1 kV/cm. Deviations from this normal situation can often be found, depending on other parameters such as surface tension and conductivity. The application of voltages up to 100 kV on electrode distances of <15 cm and >20 cm can sometimes also be the right setting. In cases of low-vapor-pressure solvents, small electrode distances do not allow complete solvent evaporations before reaching the collector and sticky fiber mats or thick fibers are obtained. Worth mentioning here is a special variation of electrospinning with electrode distances of less than 1 cm, called *precision or near-field* electrospinning, which is useful for making targeted structures and patterns (comparable to 3D-printing). The deposition of electrospun fibers in conventional electrospinning is not localized due to bending instabilities, giving looping motions to the jet. By reducing the distance between the electrodes in precision or near-field electrospinning, the bending instabilities are suppressed and fiber deposition is mainly controlled by the straight path of the jet, providing predesigned patterns [1].

Another important processing parameter to be adjusted while electrospinning is the rate at which the polymer solution is fed to the nozzle (flow rate). It can vary anywhere between 0.01 and 1 ml/min for a nozzle with an inner diameter of 0.3 mm. As a general rule, flow rates should be low for getting thin fibers. Flowing polymer solutions at high rates lead to thicker, beaded fibers. It is always necessary to adjust the flow rate for a given voltage and other spinning parameters, in order to get a stable Taylor cone during the electrospinning process. Flow rates below and above this threshold value provide droplets, beaded, splitted and/or thick fibers.

The surface tension, the conductivity, and the dielectric permittivity are more or less controlled by the solvent used for electrospinning. Water is very often employed as a solvent for electrospinning. It has a surface tension of 72 mN/m, a conductivity of 0.45 mS/m and a dielectric permittivity of 89 F/m at 0 °C. Organic solvents with conductivities of 0.05–30 mS/m and surface tensions of 20–75 mN/m are also suitable for electrospinning. The most common organic solvents utilized for electrospinning are THF, DMF, dichloromethane, chloroform, acetone, isopropanol, etc.

After providing a brief description of parameters that affect the fiber formation, this chapter helps learn to make nanofibers by electrospinning and to characterize them in the coming sections. Representative examples of the preparation of electrospun monolith fibers of water-soluble polymer, biodegradable polymer, polymer dispersion, high-performance polyimide, composite fibers with metal nanoparticles, bicomponent fibers with side-by-side and coaxial morphology and 3D tubular structures are described in this chapter.

4.2 Electrospinning of synthetic water-soluble polymers

Poly(ethylene oxide) (PEO), poly(vinyl alcohol) (PVOH), and poly(vinyl pyrrolidone) are some of the most common examples of synthetic water-soluble polymers, which have been electrospun with or without additives such as enzymes and drugs for various applications in the literature. The electrospinning procedures for nanofibers of water-soluble polymers, taking PVOH as a representative example, by a single-nozzle/syringe electrospinning are described next.

4.2.1 Electrospinning of poly(vinyl alcohol): Randomly oriented and aligned fibers

PVOH is a water-soluble, semicrystalline, and biocompatible polymer, which is used for various applications in cosmetic, medical, pharmaceutical, paper, adhesive, coating and packaging industries. It is made by a basic hydrolysis of poly(vinyl acetate) (PVAc) (Scheme 4.1) with sodium hydroxide, since the vinyl alcohol monomer for a direct polymerization is not available due to tautomerism to acetaldehyde.

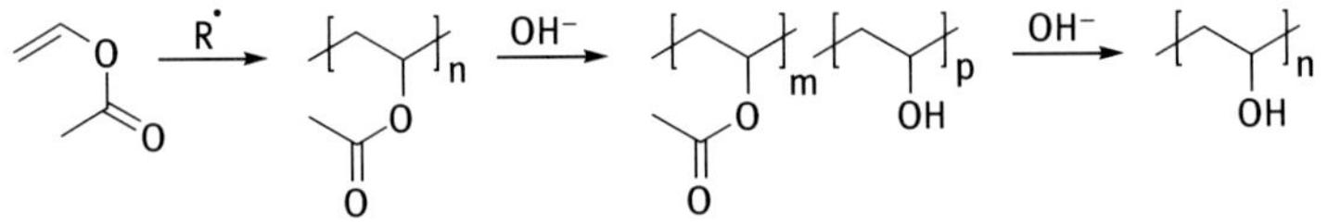

Vinyl acetate   Poly(vinyl acetate)   Poly(vinyl acetate)   Poly(vinyl alcohol)
                                            -co-
                                            Poly(vinyl alcohol)

Scheme 4.1: Synthesis of poly(vinyl alcohol). Poly(vinyl acetate), the precursor polymer, is made by radical polymerization and hydrolyzed to poly(vinyl alcohol).

PVOH is commercially available in different grades, differing not only in molar mass but also in the hydrolysis degree (i.e., the fully hydrolyzed and the partially hydrolyzed PVOH have different amounts of acetate groups in the backbone). The solubility of PVOH in water is affected by the degree of hydrolysis. PVOH grades with higher degrees of hydrolysis (98–100%) require either longer times or higher temperatures (90 °C) for dissolution in water due to an increased crystallinity. The glass transition temperature (T_g) ranges from 40 °C to 80 °C, and the melting temperature (T_m) from 180 °C to 240 °C, depending on the hydrolysis grade and the molecular weight. PVOH is also soluble in organic solvents like DMSO and DMF. The commercially available grades of PVOH (Mowiol) are marked with two numbers, i.e., Mowiol 15-79, Mowiol 3-96 and Mowiol 10-98. The first number represents the viscosity (mPa s) of a 4% aqueous solution at 20 °C, measured according to the specifications mentioned in the DIN 53015. The second number illustrates the average degree of hydrolysis, which is also called saponification (in mol%).

PVOH (87–89% hydrolyzed; M_w (weight average molar mass): 146,000–186,000 g/mol
Sigma-Aldrich; CAS-Number: 9002-89-5); distilled water; doctor's syringe: 1 ml (outer
diameter of syringe needle: 0.6 mm).

Preparation of PVOH solution for electrospinning
Weight percentages of 5–12% are spinnable, but one needs a lower relative humidity
for electrospinning of solutions with lower concentrations. Prepare, for example, a
7.6 wt% solution by adding 0.76 g of the PVOH to 9.24 g of distilled water while stirring. For dissolution, heat the suspension in a water bath at 90 °C in a closed bottle,
in order to retain the targeted concentration, for about 24 h. If the solution contains
air bubbles, one can use a shaker to remove them. Inhomogeneous electrospinning
solutions will not spin properly, since the jet will break up and droplets on the collector will be the consequence.

Preparation for electrospinning
Take out one needle (diameter: 0.6 mm), cut off the sharp needle tip and smoothen
the needle tip with an abrasive paper until the tip becomes flat. Blunt-tip/flat-tip
needles can also be bought and directly employed. Suck 0.5 ml of the solution,
made as described earlier in a 1-ml doctor's syringe. Fix the blunt-tip needle to the
syringe. Mount the filled syringe (with the PVOH solution) to a syringe pump. Note:
Higher-volume syringes with more polymer solution can also be taken.

Electrospinning Equipment
A single-nozzle apparatus was used for the present experiment. There are many
companies worldwide that are selling single-nozzle electrospinning machines. One
can get more information on the Internet. The major components of an electrospinning machine, as described in Chapter 2, can also be self-assembled in own univer-
sity/company workshops.

Attention! Use secure machines as high voltage is applied.   ⚡

> **!** **Note.** The electrospinning procedures in this book are given based on a house-made (university work-shop) machine. Fine-tuning of the conditions might be required, depending on the apparatus used for electrospinning. Flow rate, voltage, humidity, and electrode distance, for example, need to be adjusted accordingly in order to get fibers.

Electrospinning procedure for making randomly oriented fibers

Turn on a table lamp placed very close to the spinning apparatus to help with the observation of the fibers during electrospinning. Place a stationary collector beneath the needle at a distance of about 20 cm. Aluminum foil or glass plates are usually used as collector; other types of collectors can also be used. Set the humidity to 20% RH. The temperature should not be lower than 20 °C, since the vapor pressure of water is not very high. Adjust the flow rate to 0.4 ml/h and turn on the voltage, which is about 18 kV for the syringe and 0.1 kV for the collector. If the deposition of the fibers is too indefinite, one can increase the voltage of the collector up to 2 kV, but too high counter voltages generate uneven fibers.

Characterization

Before electrospinning, the solution was characterized for viscosity, surface tension, and conductivity. The rheological measurements were performed on an Anton Paar MCR500 with a cone–plate setup (diameter: 50 mm; opening angle: 1°; gap height: 51 μm). The specimen was tempered at 293 K with a TEK 150P-C Peltier system. The surface tension was measured on a Dataphysics DCAT 11 machine, equipped with a standard Wilhelmy plate (PT11). The conductivity was quantified with an Inolab Term-inal 3 measuring system. The spinning solution had a conductivity of 506 ± 0.1 μS/cm and a surface tension of 47.77 ± 0.03 mN/m at 296 K. The rheological measurements are depicted in Figures 4.1 and 4.2, and the viscosity values are given in Table 4.1. The zero shear viscosity (η_0) of the controlled shear rate test and the frequency sweep test ($\lim_{\omega \to 0} \eta^*(\omega)$) are almost the same, which implies a validity of the Cox–Merz rule (see also "Interesting to know section"). Figure 4.2 shows the values for G'' and G', in which the values of G'' are always higher than the values of G'. This stands for a purely liquid behavior of the spinning solution, i.e., no buildup of a network.

Table 4.1: Results of rheological measurements on PVOH solution.

Unit	Result (at 293 K)
η_0	738 mPa s
$\lim_{\omega \to 0} \eta^*(\omega)$	797 mPa s

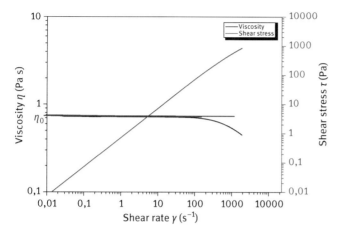

Figure 4.1: Controlled shear rate test of a PVOH solution with 7.6 wt% concentration.

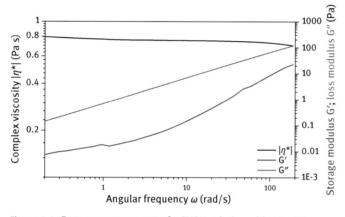

Figure 4.2: Frequency sweep test of a PVOH solution with 7.6 wt%.

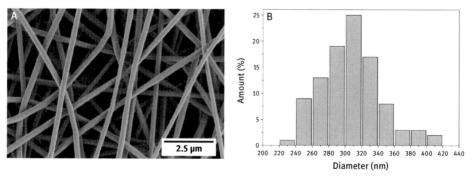

Figure 4.3: Fiber morphology of randomly oriented electrospun PVOH fibers as observed by a scanning electron microscope (A) and fiber diameter distribution (B). The average fiber diameter is 308 ± 37 nm. A minimum of 100 fibers should be checked for calculating the average fiber diameter.

The fiber morphology and average diameters as observed by a scanning electron microscope (SEM, Zeiss 1530 with 1 kV acceleration voltage and a secondary electron detector) are shown in Figure 4.3.

Aligned PVOH fibers

The fibers with different degree of alignments can be obtained by collecting them on a rotating disk or a drum. The speed of rotation of the drum/disk controls the degree of alignment of the fibers. The words *orientation* and *alignment* are often used interchangeably in the literature related to the electrospun fibers. They both should be differentiated. Alignment could be referred to the macroscopic arrangement of individual fibers in a nonwoven. The molecular orientation of macromolecular chains within a fiber is termed as *fiber orientation*.

The high elongation and elongation rates during electrospinning, together with polymer structure with slow relaxation, provide molecular orientation. Crystalline polymers such as poly(ethylene oxide) and poly(oxymethylene) show molecular orientation in electrospun fibers most probably due to the very fast crystallization process during electrospinning [2]. The crystals hinder the macromolecular chain relaxation in electrospun fibers after fiber formation, leading to molecular orientations. Amorphous polymers with high glass transition temperatures can also show molecular orientation as extended macromolecular chains after electrospinning lack sufficient mobility to relax at room temperature. Molecular orientation can affect crystallinity, crystallite size, melting enthalpy, secondary structure, and mechanical properties. Polarized IR is one of the easiest ways for checking the macromolecular orientation within fibers. The polarized spectra overlap for random orientation, whereas they show significantly different absorbance for fibers with macromolecular orientations.

Here we describe the formation of aligned PVOH fibers using three syringes (spinning at the same time) and collecting fibers on a metal drum (wrapped with backing paper), with a length of 45 cm and a diameter of 15.5 cm, rotating at a speed of 55 km/h (1,270 rpm; Figure 4.4). For the power connection of the three syringes, the needles were connected with each other with a metal wire and the power supply was connected to the middle syringe. The similar procedure can also be used with a one-syringe electrospinning apparatus.

The voltage used for electrospinning was +19.5 kV at the syringes and −1 kV at the rotating collector. The flow rate of the PVOH solution was 371 µl/h, and it took 8 h to get a 14-µm thick fiber mat, which can be easily peeled off.

The aligned fibers can also be collected on a metal frame as shown in Figure 4.5.

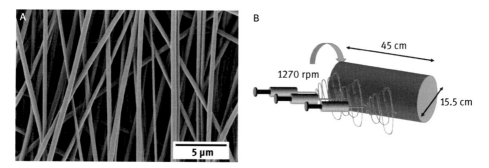

Figure 4.4: SEM image of electrospun aligned PVOH fibers (A) and setup for the production of the aligned fibers (B).

Figure 4.5: Aligned fibers collected on a metal frame.

The diameter distribution and the angle of fiber alignment as determined by SEM are shown in Figure 4.6. The mean diameter of the fibers is 411 ± 86 nm. The alignment of fibers can be controlled by adjusting the speed of the rotation of the collector. The degree of fiber alignment S can be determined using Equation (4.1). Θ is the angle a fiber makes with respect to the direction of fiber alignment. Θ will be 0 for perfectly aligned fibers.

$$S = \frac{<3\cos(\theta)^2 - 1>}{2} \tag{4.1}$$

The value of S varies between 0 and 1 ; 0 for randomly aligned and 1 for perfectly aligned fibers.

For many areas of the use of electrospun nonwovens such as filter applications and tissue engineering, it is interesting to know the pore size as it decides the efficiency of filtration and cell proliferation and growth. Mercury intrusion porosimeter, liquid extrusion porosimeter and capillary flow porometer are mostly used for measuring the pore size and pore size distribution. Here we describe the general procedure for measuring the pore size using a capillary flow porometer of the type TOPAS PSM 165 with a sample diameter of 11 mm and test fluid Topor (perfluoro compound) as carried out on PVOH fiber mat.

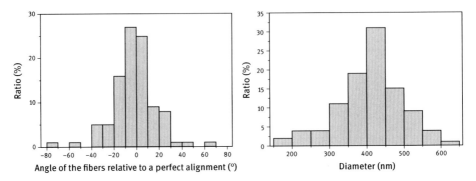

Figure 4.6: Relative angles and diameters of the aligned PVOH fibers.

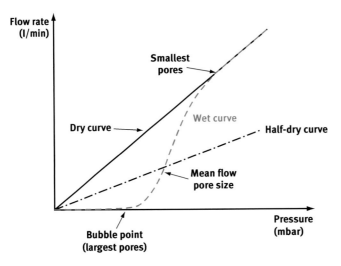

Figure 4.7: Pressure vs. flow rate curves for the determination of pore size.

In general, an inert/non-reacting gas such as nitrogen or air is passed through the sample and a change in flow rate as a function of pressure is monitored. The measurement is done on both dry and wet sample in which the pores are intentionally closed with a wetting liquid like Topor. The airflow and pressure is related to the pore size, and the larger pores are opened first followed by smaller pores on increasing the gas pressure. The opening of pores is monitored by observing continuous evolution of gas bubbles in a liquid column kept over the sample. The pressure required for opening the largest pore size, i.e., for making the wetted sample permeable, is designated as the first bubble point (Figure 4.7). The pore size is calculated according to the Young–Laplace formula $P = 4 \times \cos(\theta) \times y/D$; P: the pressure necessary for liquid displacement from the pore, θ: the contact angle of the wetting fluid with the material, y: surface tension of the wetting liquid and D: the pore diameter. Mean flow pore size signifies the pore size at which there is a 50% pressure drop and is taken as the size at which 50% pressure drop between a wet and a dry sample occurs.

A pore size test on the oriented PVOH fiber mat prepared earlier gave an average bubble point (opening of the largest pores) of 1.23 ± 0.04 µm and a mean flow pore size of 1.03 ± 0.03 µm. The pore size is dependent on the fiber diameter, length and mass and plays an important role in filter applications as described in detail in Chapter 5.

One of the important properties of electrospun nonwovens is high porosity required for applications such as catalysis and drug release. Porosity (ϵ) is an indication of free space (pores between the fibers and different fibrous layers). It can be calculated using the density (ρ) of the fiber mat and the corresponding bulk material (Equation 4.2). The density of the fiber mat can be easily determined by dividing the mass of the electrospun nonwoven mat of a particular volume (length × width × thickness). The porosity calculated for the aligned PVOH fiber mat prepared earlier was ca. 92%.

$$\epsilon = 100\% - \frac{\rho_{\text{fiber mat}} \cdot 100\%}{\rho_{\text{bulk material}}} \tag{4.2}$$

ϵ porosity (free space) of the fiber mat
ρ density (PVOH bulk material: 1.269 g/cm³)

Further characterization

The fiber mats can be characterized further, for example, for thermal stability, phase transitions and molecular characterization depending on the application and need. The structural characterization or the presence of the respective polymers in a fiber mat with more than one component can be easily done by IR measurements in solid state. The IR spectrum of PVOH fiber mat as measured with a Digilab Excalibur 3000 FT-IR with a resolution of 4 cm^{-1} and an ATR unit is shown in Figure 4.8. The prominent peaks were observed at $\tilde{v}$ (cm^{-1}): 3314 (m, v OH–H); 2936 (m, v_{as} CH$_2$); 2913 (m, v_s CH$_2$); 1732 (m, v C=O); 1246 (s, v C–C), 1092 (s, v C–OH). The presence of C=O peak at around 1732 cm^{-1} originates from acetate functional groups as PVOH is made by hydrolysis of poly(vinyl acetate) and can contain different proportions of acetate groups depending on the grade used for electrospinning.

4.3 Biodegradable and biocompatible fibers

Biodegradable and biocompatible polymeric nonwovens are promising candidates as carriers for drugs, enzymes, cells, and bacteria for a controlled drug release, catalysis, wound dressings, and bioreactions. They are also useful as scaffolds for tissue engineering, mimicking the extracellular matrix. Both natural and synthetic biodegradable polymers have been electrospun in the literature. The examples include polycaprolactone, poly(D,L-lactide), poly(L-lactide) and their copolymers, poly(ester urethane), poly(hydroxyl butyrate), cellulose and derivatives of cellulose, silk fibroin (SF), gelatin, chitin, fibrin and fibrinogen. As an example, the electrospinning of poly(L-lactide) is described in the following section.

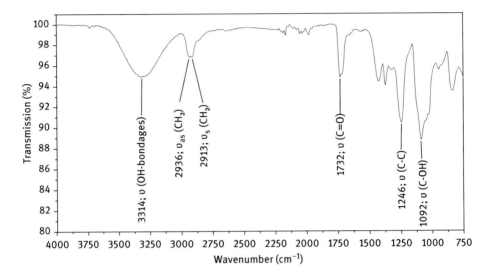

Figure 4.8: FT-IR spectrum of PVOH fiber mat.

4.3.1 Poly(L-lactide) nanofibers: effect of concentration and conductivity on fiber formation

Poly(L-lactide) (PLLA) is a synthetic biobased, biodegradable and semicrystalline polymer with a glass transition temperature of around 60 °C and a melting point of ca. 160 °C. The technical synthesis involves a metal-catalyzed ring-opening polymerization of L-lactide, using a tinoctoate catalyst (Scheme 4.2).

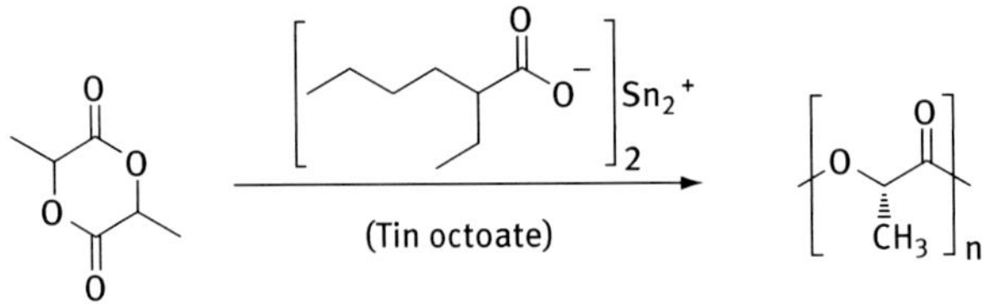

Scheme 4.2: Synthesis of PLLA via a metal-catalyzed ring-opening polymerization.

Materials

PLLA (M_w = 670,000; M_w/M_n = 1.60, CAS number: 33135-50-1; L 210 S; Boehringer/
Ingelheim), dichloromethane (technical grade, distilled before use).

Fiber formation by electrospinning

Make a 3 wt% solution in dichloromethane by dissolving 0.3 g PLLA in 9.7 g dichloromethane at room temperature. The single-nozzle electrospinning apparatus was used
as described in Section 3.2.1. An electrode distance of 14 cm, a flow rate of 1.3 ml/h,
room temperature of ca. 20 °C and a voltage of 40 kV (at the syringe) gives PLLA fibers
with diameters of 300–800 nm. In this setup the collector was grounded.

Increasing the polymer solution concentration to 5 wt% still gives continuous and
smooth fibers but with bigger diameters (1–2 μm) (Figure 4.9). Dilute solutions (1–2 wt%)
give beaded fibers, and no fiber formation will be observed for a solution <1 wt%.

Attention! The choice of solvent is highly important for getting a stable electrospinning jet. Other
organic solvents or a mixture of solvents like chloroform–DMSO can also be used for electrospinning
of PLLA, but the concentration range required for getting smooth fibers will be different.

Figure 4.9: Effect of the solution concentration on the fiber formation by electrospinning: PLLA electrospun from dichloromethane using (A) 5 wt% (B) 4 wt% and (C) 1 wt% solutions.
[Reprinted from Z. Jun, H. Hou, A. Schaper, J. H. Wendorff, A. Greiner, Poly-L-lactide nanofibers by electrospinning–Influence of solution viscosity and electrical conductivity on fiber diameter and fiber morphology, e-Polymers **2003**, *009*.]

In addition, the conductivity of the polymer solution can influence the spinning
process to a large extent. Sometimes an increase in solution conductivity can lead to a
stable spinning jet with the formation of continuous smooth fibers without beads. The
conductivity of a 2 wt% solution of PLLA in dichloromethane changes from 0.043 μS/cm
(without salt) to 1.7–3.6 μS/cm [different amounts of pyridinium formate(PF) added
(0.2–0.8 wt%)], leading to changes in the fiber morphology (Figure 4.10) from
beaded fibers to smooth fibers, without affecting the solution viscosity.

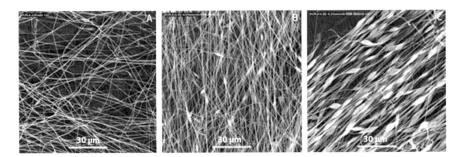

Figure 4.10: SEM photographs of electrospun fibers obtained from solutions of 2 wt% PLLA in dichloromethane and varying amounts of PF. (A) 0.8%, (B) 0.5%, (C) 0.2%. [Reprinted from Z. Jun, H. Hou, A. Schaper, J. H. Wendorff, A. Greiner, Poly-L-lactide nanofibers by electrospinning–Influence of solution viscosity and electrical conductivity on fiber diameter and fiber morphology, *e-Polymers* **2003**, *009*.]

The addition of salt even provides fibers from otherwise nonspinnable concentrations. A very dilute (0.8 wt%) PLLA solution in dichloromethane is not spinnable, but an addition of 0.8 wt% PF (referred to the solvent) increases the solution conductivity to 1.26 µS/cm and provides 10–70 nm thin fibers (Figure 4.11). Note that the effect of the addition of salt on the fiber diameter and solution viscosities can vary from system to system. Charged polymers might be affected in a different way.

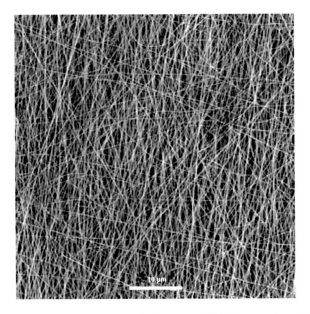

Figure 4.11: SEM picture of electrospun PLLA fibers, using a 0.8 wt% PLLA solution in dichloromethane with 0.8 wt% PF. [Reprinted from Z. Jun, H. Hou, A. Schaper, J. H. Wendorff, A. Greiner, Poly-L-lactide nanofibers by electrospinning–Influence of solution viscosity and electrical conductivity on fiber diameter and fiber morphology, *e-Polymers* **2003**, *009*.]

If you prepare a polymer solution of an appropriate concentration, which provides stable electrospinning jets and smooth fibers, an addition of salt could change the average fiber diameter without disturbing the spinning process. For example, the addition of 0.8 wt% PF to a 5 wt% PLLA solution in dichloromethane provides thinner fibers (Figure 4.12).

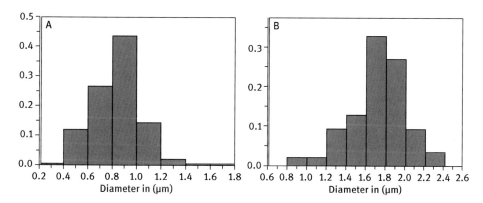

Figure 4.12: Diameter distribution of PLLA fibers, fabricated from a 5 wt% PLA solution in dichloromethane with 0.8 wt% PF (A) and without PF (B). The amount of PF is referred to the weight of dichloromethane. [Reprinted from Z. Jun, H. Hou, A. Schaper, J. H. Wendorff, A. Greiner, Poly-L-lactide nanofibers by electrospinning–Influence of solution viscosity and electrical conductivity on fiber diameter and fiber morphology, e-Polymers **2003**, *009*.]

4.4 Silk fibroin fiber formation by electrospinning

A large number of naturally occurring polymers such as cellulose derivatives, SF, collagen, gelatin, hyaluronic acid, fibrin, fibrinogen and elastin can be spun into fibers using electrospinning. Silk is one of the naturally occurring proteins with interesting properties such as biocompatibility, biodegradability and superior mechanical properties. It is used as biomaterial for various applications in the form of films, sponges, gels, particles and fibers. It is produced either by silk worms (e.g., *Bombyx mori*) or web-weaving spiders like *Araneus diadematus* or *Nephila clavipes*. Details about the silk structure are given in Chapter 1. The *B. mori* cocoon contains sericin, a glycoprotein covering fibroin fibers. Therefore, the sericin part has to be removed before processing SF. The separation procedure of SF from sericin and processing in the form of fibers using electrospinning is described here.

Materials

B. mori silk was bought from SeidenTraum, Germany. 1,1,1,3,3,3–Hexafluoro-2-propanol (HFIP, > 99.9%) was obtained from Apollo Scientific Limited, UK.

Preparation of SF spinning solution

Sericin was removed by treating the *B.mori* silk with NaHCO$_3$ solution. 15 g of *B.mori* silk is heated at 90 °C for 1.5 h in 3L NaHCO$_3$–H$_2$O (0.5 wt%). After this the insoluble SF is taken out and washed with water. The degumming process with NaHCO$_3$–H$_2$O was repeated once more. The degummed silk was finally washed with H$_2$O three times and dried in an oven at 90 °C overnight. 70 mg dried degummed silk was dissolved in 1 ml HFIP by stirring for about 4 weeks. The solution was centrifuged and insoluble flakes were separated, and the clear solution was used for electrospinning. The final concentration of SF was about 100 mg/mL. The conductivity of the solution was 20.3 µS/cm as measured by InoLab Terminal 3 instrument (WTW GmbH, Germany).

Electrospinning

The following spinning parameters were used on a single-nozzle electrospinning apparatus: +9 kV at the needle (diameter of needle 1.2 mm) and –1 kV at the counterelectrode. The spinning was done at room temperature (20 °C) using a flow rate of 0.26 ml/h. The fiber morphology as seen by SEM is shown in Figure 4.13.

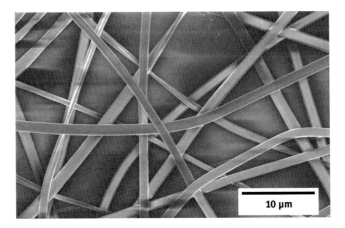

Figure 4.13: SF electrospun fibers.

Under given spinning conditions, flat ribbon-shaped SF fibers were observed. The fast evaporation of the solvent leads to flattening of fibers due to the skin formation. A change in spinning conditions such as type and mixture of solvents and conductivity can provide morphologies different from the one given in Figure 4.13. The secondary structure of SF fibers can be analyzed by FT-IR. Further details about the secondary structure and the use of SF in making 1D biocomposite fibers can be referred in the literature [3].

4.5 Aqueous dispersion electrospinning: Green electrospinning

In general, the process of solution electrospinning uses a large amount of solvents. The spinnable polymer concentration in conventional electrospinning is in the range of 1–20 wt%. This means that 99–80 wt% is the solvent in spinning formulations. Although there are few polymers that are soluble in water and can be spun from nontoxic, environmental-friendly solvent water, many other technically useful polymers such as polyamides, polyesters, polycarbonates and polyimide precursors are water-insoluble polymers and require toxic and flammable organic solvents for electrospinning. It would be preferable to use water as solvent for making fibers for applications in the biomedical field, the foodstuff industry, the agricultural field and many others. Moreover, to make the spinning process safe during large-scale fiber production, it would be desirable to avoid the use of flammable organic solvents. The only alternative available at the moment is the spinning of primary and secondary aqueous dispersions of otherwise water-insoluble polymers. The polymers can be dispersed in the form of nano-micro particles in water stabilized by surfactants. By using special polymers, even the use of surfactants in polymeric dispersions can be avoided. The electrospinning of aqueous dispersions without additional surfactants is named as *green electrospinning*. More details about this topic are provided in Chapter 5. Here we provide the procedure for electrospinning of one of such aqueous dispersions [4].

4.5.1 Biodegradable fibers by aqueous dispersion electrospinning

The biodegradable fibers based on aliphatic polyesters can be spun easily from organic solvents. One of the examples is given in Section 4.2 regarding the spinning of PLLA fibers from organic solvents. For using water as a solvent for spinning, the corresponding water-insoluble polymers have to be dispersed in water (secondary dispersion). Various state-of-the art methods such as the solvent displacement method and nanoprecipitation can be used for making polymeric secondary dispersions. The aqueous polymeric dispersions can also be made by direct polymerization in water (primary dispersions) (Chapter 5). Generally, various surfactants are used to provide stability to the polymeric dispersion. Amphiphilic biodegradable polymers with hydrophilic and hydrophobic blocks can be dispersed in water without the need of any additional surfactant. In the following section we describe the formation of fibers from aqueous dispersion of one such polymer, namely polycaprolactone-block-poly(ethylene glycol) (PCL-MPEG). The block copolymer system of poly(ethylene glycol) (PEG) and poly(ε-caprolactone) (PCL) is of special interest since PEG is biocompatible and offers a "stealth" behavior *in vivo* due to its ability to minimize cell and protein interactions, whereas PCL is both biocompatible and biodegradable due to the enzymatic and/or pH-dependent breakdown of ester bonds. That is why electrospun nanofibers that are fabricated by electrospinning of aqueous dispersion of PCL-b-MPEG can be applied for biomedical and agricultural applications. The block copolymer forms micellar particles due to the amphiphilic affinity toward water. Here, the MPEG block has a high affinity for water, whereas the PCL block is insoluble in water. Even the polymeric dispersions with a very high solid content (ca. 40 wt%) have a very low viscosity and therefore require a small amount of template polymer (any water-soluble polymer) in the spinning formulation to provide appropriate viscosity for spinning. The template polymer can be easily removed after fiber formation by simply putting fibers in water.

Materials

α-Hydroxy-ω-methoxy-PEG (MPEG; CAS: 9004-74-4; Sigma-Aldrich; dried at 40 °C under vacuum; M_n = 5,000 g/mol), ε-caprolactone (CL; CAS: 502-44-3; Alfa Aesar, 99% pure; dried over CaH_2 and freshly distilled under vacuum) and tin octoate (stannous octoate; CAS: 301-10-0; Sigma-Aldrich) were used for making polymers and fibers. PCL-b-MPEG was synthesized in a flame-dried and argon-purged flask in laboratory using the following procedure (Scheme 4.3):

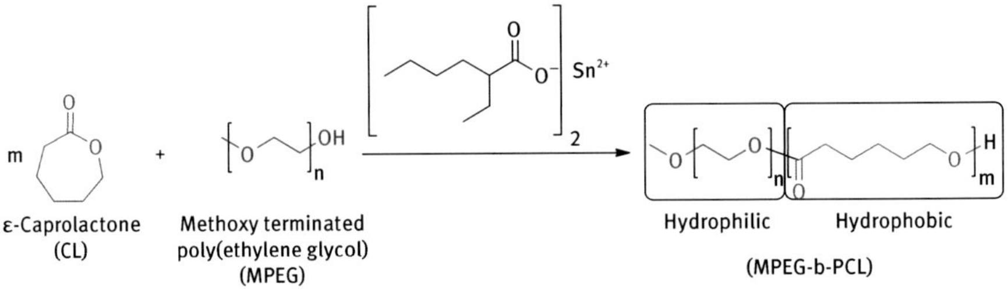

Scheme 4.3: Synthesis of MPEG-b-PCL using stannous octoate as catalyst.

The reactants 5 g of MPEG, 15 g of ε-caprolactone, and 10 μl of tin octoate catalyst were added in a dried flask. Afterward, the reaction mixture is heated to 110 °C for 16 h. Subsequently, the product is purified by dissolving the solid reaction mixture in chloroform, precipitating in n-pentane, filtering, extracting with water and freeze-drying. The molar mass (M_n; number average molar mass) of the PCL block was 15,000 g/mol as determined by gel permeation chromatography using THF as eluting solvent.

Preparation of dispersion of PCL-b-MPEG in water

For making a secondary dispersion of PCL-b-MPEG in water, 1.5 g of the polymer (PCL-b-MPEG) was mixed with 8.25 ml of deionized water and heated for 15 min at 90 °C while stirring. Afterward, add 0.25 g of PEO (900,000 g/mol; CAS: 25322-68-3) to the dispersion and heat with stirring again for 15 min at 90 °C in a hot water bath. For complete dissolution of PEO, the dispersion is stirred further overnight at room temperature and a white, highly viscous dispersion is provided (Figure 4.14). The final dispersion contains 15 wt% of the PCL-b-MPEG and 2.5 wt% of PEO as a template polymer. The size of micellar particles was measured by dynamic light scattering (DLS) (3D LS Spectrometer of LS Instrument AG (Fribourg Switzerland); HeNe laser, wavelength: 632.8 nm; refraction index: 1.3320, viscosity: 0.899 mPa s; concentration: 0.01 wt%, angle: 90°) and was found to be 52 ± 0.26 nm and had a monomodal dispersity. The surface tension as determined by Dataphysics DCAT 11 machine at 295 K, equipped with a standard Wilhelmy plate (PT11),was 60.15 ± 0.03 mN/m. The conductivity of 132.9 µs/cm was measured at 22 °C with an Inolab Terminal 3 measuring system.

Figure 4.14: PCL-b-MPEG aqueous dispersion used for the organic solvent–free electrospinning approach (green electrospinning).

Electrospinning of PCL-b-MPEG dispersion

One milliliter of the dispersion was filled in a 2.0 ml syringe. The single-needle electrospinning apparatus was used with the following spinning parameters: outer diameter of the blunt needle: 0.6 mm; applied voltage to the syringe: 30 kV; applied voltage to the collector: grounded; collector: Al foil; flow rate: 4.2 ml/h; collecting distance: 20 cm; relative humidity: 20%. Fibers with an average diameter of 429 ± 70 nm were observed (Figures 4.15 and 4.16).

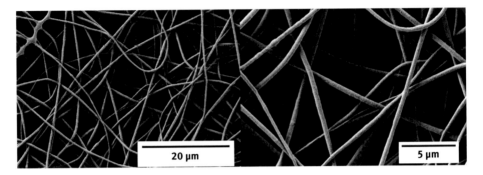

Figure 4.15: SEM images of nanofibers obtained from PCL-b-MPEG dispersion.

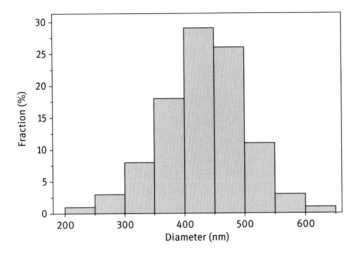

Figure 4.16: Diameter distribution of the obtained PCL-b-MPEG fibers via SEM. For the measurement, 100 fibers were tested.

4.6 Formation of high-performance polyimide fibers

Polyimides (PIs) are high-performance polymers with excellent mechanical properties, thermal stability and chemical resistance. The detailed chemistry and applications of PI are given in Chapter 5. PIs are made in two steps. The first step is the formation of polyamic acid, which undergoes imidization in the second step by elimination of water, achieved by heating at high temperatures. The PIs are insoluble in common organic solvents, and therefore they are processed using the intermediate polyamic acid. Kapton is one of the commercially available and highly studied polyimides made by polycondensation of a diamine (4,4′-oxydianiline) (ODA) and a dianhydride (pyromellitic dianhydride) (PMDA) (Scheme 4.4). The Kapton fiber formation by electrospinning is described in the following section.

4.6.1 Polyimide nanofibers by electrospinning

Materials

Pyromellitic dianhydride (PMDA, 99%, Acros Organics), 4,4′-oxydianiline (ODA, 98%, Acros Organics), DMF (99.99%, Fisher Chemical, density: 0.948 g/cm³, dried over CaH₂).

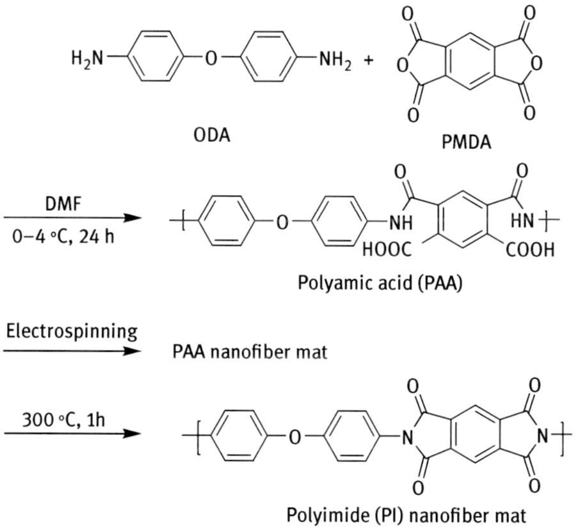

Scheme 4.4: Synthetic scheme for the formation of polyimide.

Preparation of the polyamic acid solution and electrospun PAA fibers

All working steps were performed under argon; 10.22 g of ODA (51.04 mmol, 1.01 eq.) was dissolved in 40 mL of DMF at 0 °C. Afterward, a suspension of 11.02 g of PMDA (50.52 mmol, 0.99 eq.) in 30 ml DMF was added dropwise to the previously made ODA/DMF solution. The mixture was stirred for 1 h in an ice–water bath and 5 h at room temperature. A viscous and yellow PAA solution (24 wt%) was obtained. For getting high molar mass polymers 1:1 stoichiometry between the reactants should be used with purified monomers.

The PAA solution was characterized for conductivity, viscosity, and surface tension before electrospinning. The surface tension was measured by a Dataphysics DCAT 11 surface tension machine at 23 °C. The viscosity measurements were performed on a HAAKE PK 100. The sample height was set to 200 μm, the rotation speed was 1,200 rpm and the whole test lasted for 120 s. The conductivity was measured by an Inolab Terminal level 3 at 22 °C. The values are given in Table 4.2.

Table 4.2: Physical properties of the PAA solution

Infinite shear viscosity η_∞	1.5 ± 0.3 Pa s
Conductivity	50 ± 0.2 µs/cm
Surface tension	35.19 ± 0.03 mN/m

For electrospinning, 1 ml of the PAA solution in DMF (24 wt%, synthesized earlier) was filled in a 2.0 ml syringe. The electrospinning parameters are outer diameter of the blunt needle: 0.6 mm; applied voltage to the syringe: 20 kV; applied voltage to the collector: grounded; flow rate: 0.3 ml/h; collecting distance: 20 cm; collector: Al foil.

Imidization of PAA fibers to get PI fibers

The obtained PAA nanofibers can be imidized to get PI nanofibers by thermal treatment. The fiber mat needs to be dried before imidization, which can be done at 100 °C, overnight and under vacuum. Imidization was accomplished by heating of the PAA nanofiber mat at 250 °C for 3 h under vacuum or nitrogen atmosphere to prevent the fibers from degradation with oxygen. After imidization, the color of the mat changes from pale to yellow and the fibers are no more soluble in any organic solvent. The whole process for the preparation of PI nanofibers is shown in Scheme 4.4.

The PAA and PI fiber mats were structurally characterized by FT-IR (Figure 4.17). The FT-IR spectra were recorded by a Digilab Excalibur Series with a PIKE MIRacle ATR unit and a resolution of 4 cm^{-1}. By comparison of the IR spectra of PAA and PI, it is apparent that the asymmetric and symmetric valence vibration of the amide NH group (3279 cm^{-1}) vanishes, whereas a new vibration for the imide groups appears at 1775 cm^{-1} after imidization.

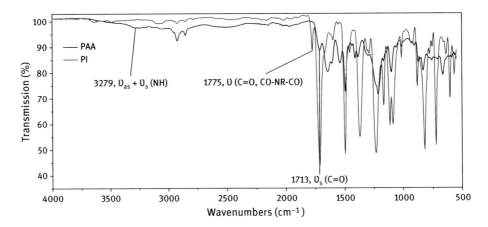

Figure 4.17: FT-IR spectra of PAA and PI nanofiber mats.

Scanning electron microscopy (Figure 4.18) (SEM; Zeiss LEO1530; acceleration voltage: 2 kV) was used to characterize the morphology and the diameter of the fibers. Before scanning, the samples were coated with 1.3 nm of platinum. The surface morphology and diameter remained the same after imidization.

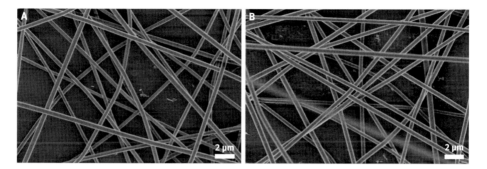

Figure 4.18: SEM images of PAA (A) and PI (B) nanofibers. The average fiber diameters are 460 ± 38 nm and 455 ± 48 nm for PAA and PI fibers, respectively. There are no changes in the surface morphology and fiber diameter on imidization.

4.6.2   Composite polymide (PI) fibers: Incorporation of antibacterial silver nanoparticles in PI fibers by electrospinning

One of the advantages of electrospinning is the easy incorporation of functional additives in fibers by simply mixing the corresponding additives in the spinning formulations. Composite fibers with many different types of additives such as dyes, drugs, enzymes, bacteria, virus, inorganic and metal nanoparticles and clays have already been studied. Silver nanoparticles are well known for their antibacterial properties and are easy to incorporate in electrospun polymeric fibers for providing an antibacterial functionality to the resulting fibers [5–7]. The preparation of PI–silver nanoparticles composite fibers is described in the following section.

Experimental section

The experiment is based on the PI nanofiber production as described in the previous section. There are different ways in which silver nanoparticles can be mixed in the spinning solution. One of the methods is to mix readymade silver nanoparticles of different size in spinning solution. Silver nanoparticles can also be generated *in situ* in the spinning solution. DMF was used as a spinning solvent for the preparation of PI fibers. Since DMF can reduce Ag^+ ions (Scheme 4.5), the spinning polyamic acid (PAA) solution can be directly mixed with a DMF/silver nitrate solution for the *in situ* generation of silver nanoparticles. The color of the spinning solution turns from yellow to brown over the time, which is due to the reduction of the silver ions. Further reduction will take place when the poly(amic acid) fibers are thermally treated for the imidization to PI fibers.

For the electrospinning, a fresh formulation should be utilized and mixed thoroughly with a AgNO$_3$/DMF solution. For example, mix 29 g of the PAA/DMF solution (24 wt% PAA) with 67 mg AgNO$_3$ or 133 mg AgNO$_3$ in 1 ml DMF, in order to get fibers with a 0.6 wt% or 1.3 wt% silver content, respectively.

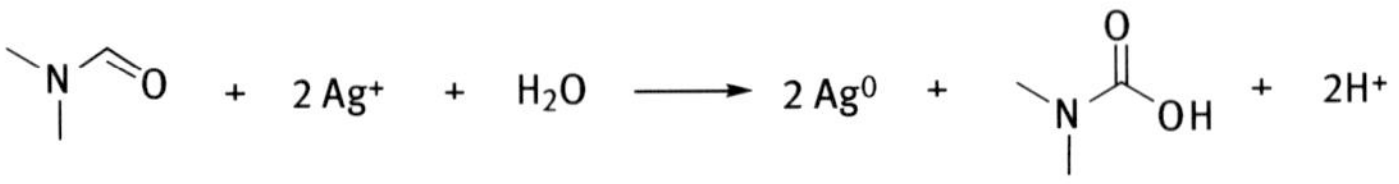

$$\text{N} \diagup\diagdown \text{O} \quad + \quad 2\,Ag^+ \quad + \quad H_2O \quad \longrightarrow \quad 2\,Ag^0 \quad + \quad \text{N} \diagdown \overset{O}{\diagup} OH \quad + \quad 2H^+$$

Scheme 4.5: Reduction of silver ions to silver via DMF.

Electrospinning conditions

For the electrospinning, a needleless electrospinning setup with a disk rotating in spinning formulation was used as one of the electrodes and the spinning was done from bottom-to-up. The voltage was applied to the disk and to an overhead hanging collector. The conditions for the electrospinning can be found in Table 4.3. Afterward, the obtained PAA nanofiber mats were imidized as described in the previous section.

Table 4.3: Electrospinning conditions for the formation of PI–silver nanoparticle composite fibers.

Sample	RH (%)	T(K)	Voltage of the disk (kV)	Voltage of the collector (kV)	Amount of fibers (g/h)
Pure PAA	24	299	46.0	−29.8	0.90
PAA, 0.6 wt% Ag	23	299	45.0	−29.8	0.44
PAA, 1.3 wt% Ag	21	298	44.5	−29.8	0.36

PAA with silver salt was spun in DMF followed by imidization to get PI–silver composite fibers.

The use of needleless electrospinning provided a broad fiber distribution between 200 nm and 1 μm. The respective fiber morphology can be seen in Figures 4.19 and 4.20. The diameter distributions of the respective silver nanoparticles are shown in Figure 4.21.

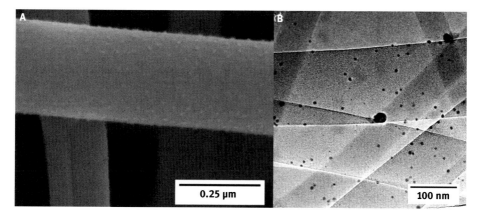

Figure 4.19: Images of electrospun PI fibers with 0.6 wt% silver content. (A) SEM image; (B) TEM image.

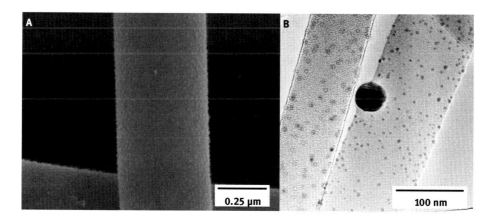

Figure 4.20: Images of electrospun PI fibers with 1.3 wt% silver content. (A) SEM image; (B) TEM image.

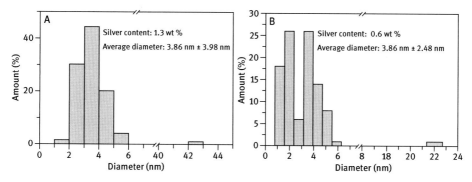

Figure 4.21: Diameter distribution of silver nanoparticles (referred to the TEM images), dispersed inside PI fibers. (A) Fibers with a silver content of 1.3 wt%; (B) Fibers with a silver content of 0.6 wt%. For averaging the particle size, 100 particles are considered.

For quantifying the antibacterial impact of the loaded PI fibers, Kirby–Bauer tests (see also 'Interesting to know' section) and serial dilution tests were performed against *Escherichia coli* (*E. coli*; Gram-negative) and *Bacillus subtilis* (Gram-positive) bacteria. These are standard test methods for checking antibacterial activity. The zone of inhibition (a clear zone showing no bacterial growth, Figures 4.22 and 4.23) increases with the increase in the amount of silver in composite fibers and pure PI fibers did not show any zone of inhibition on agar plates. The zone of inhibition is an indication of antibacterial activity due to leaching of the active agent in the composite fibers.

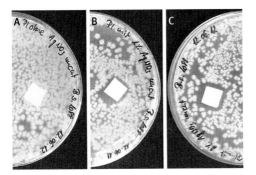

Figure 4.22: Photos of the Kirby–Bauer test on electrospun PI fibers with *B.subtilis* bacteria. (A) Without silver nanoparticles; (B) with 0.6 wt% of silver nanoparticles; (C) with 1.3 wt% of silver nanoparticles. The zone of inhibition is rising with the silver content, whereas the unloaded PI fibers show no zone of inhibition.

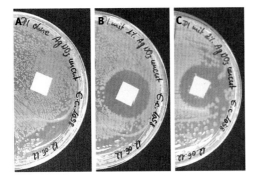

Figure 4.23: Photos of the Kirby–Bauer test on electrospun PI fibers with *Escherichiacoli* bacteria. (A) Without silver nanoparticles; (B) with 0.6 wt% of silver nanoparticles; (C) with 1.3 wt% of silver nanoparticles. The zone of inhibition is almost the same for the loaded PI fibers, whereas unloaded PI fibers show no zone of inhibition.

To quantify the antibacterial activity, the serial dilution tests are carried out (Figures 4.24 and 4.25). The test specimens were taken from the Kirby–Bauer test and washed in sterile phosphate buffer solutions. Afterward the solutions were diluted (tenfold) and again applied onto agar plates, which were incubated in turn for 24 h at 37 °C. The number of bacterial colonies was counted, and antibacterial efficiency was compared between monolith PI and PI–silver nanoparticle composite fibers, using the following equations for the percent reduction values and log reduction values, respectively:

$$\mathrm{ABE}(\%) = \frac{A-B}{A} \cdot 100$$

ABE   antibacterial efficiency (percent reduction)
A   the number of living bacteria on unloaded PI fibers
B   the number of living bacteria on PI fibers with silver nanoparticles

$$\mathrm{LR} = \log_{10}\left(\frac{A}{B}\right)$$

LR   log reduction

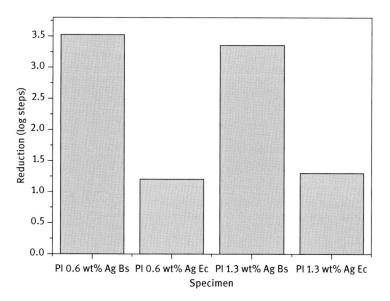

Figure 4.24: Antibacterial efficiency of PI fibers with silver nanoparticles in log steps, compared to pure PI fibers. Bs: *B.subtilis* bacteria; Ec: *E.coli* bacteria.

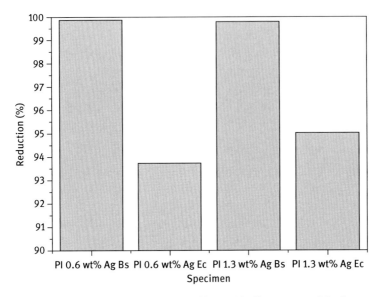

Figure 4.25: Antibacterial efficiency of PI fibers with silver nanoparticles in percentages, compared to pure PI fibers. Bs: *B.subtilis* bacteria; Ec: *E.coli* bacteria.

4.7 Special fiber arrangements via electrospinning

4.7.1 3D porous hollow tubes by electrospinning

Materials
Polycaprolactone (PCL) (Sigma-Aldrich, M_n 70,000–90,000 g/mol), formic acid (Sigma-Aldrich, 98 wt%), acetic acid (Sigma-Aldrich, 98 wt%).

Preparation of 3D tubes by direct electrospinning
A 12.5 wt% PCL solution was prepared for electrospinning by dissolving PCL in a solvent mixture of formic acid and acetic acid with a weight ratio of 1:2. The electrospinning was performed by applying a voltage of +18 kV to the needle and −6.5 kV to the collector. For making tubes by direct electrospinning, a needle with a diameter of 0.6 mm mounted on a mechanical stirrer was used as a collector. The collector was rotated at a speed of 500 rpm (Figure 4.26). The flow rate of the polymer solution was set to 1.0 ml/h. Fiber deposition was continued for about 20 min. The PCL nanofibers were deposited on the rotating needle and formed a tubular morphology. The PCL tubes can be cut into several pieces depending on the length of the tube required and can be pulled off from the needle substrate carefully. The resulting tubes had an inner diameter of ca. 0.76 mm and an outer diameter of 1.02 mm, as shown in Figure 4.27.

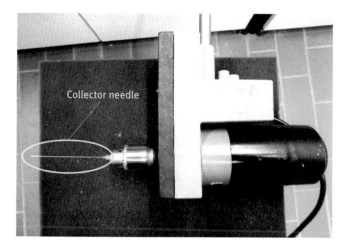

Figure 4.26: Photo of the collector setup, used for the preparation of PCL tubes.

A digital microscope (VH-Z500) and a SEM (Zeiss LEO1530, EHT = 2 kV) were used to characterize the morphology of the fibers. The sample for the digital microscopy was placed on a glass slide, while the sample for SEM was placed on an aluminum foil. Before SEM measurements, the sample was sputtered with 1.3 nm of platinum. The tubular walls were made up of aligned fibers.

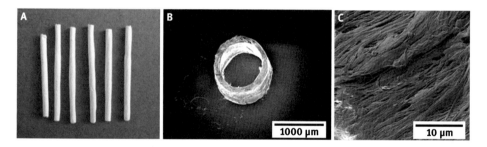

Figure 4.27: (A) Photo of the electrospun PCL tubes. SEM images of (B) sideview of the tube and (C) aligned fiber morphology on the tubular outside wall.

4.7.2 Bicomponent fibers – Coaxial and side-by-side electrospinning

Coaxial and side-by-side electrospinning are the valuable modifications of single-component electrospinning as described in Chapter 2. Coaxial spinning allows encapsulation of polymers, low-molar-mass additives, reinforcing agents, etc., in the core of a core–shell structure in which the shell can be an organic/inorganic polymeric material. Side-by-side electrospinning provides Janus fibers with two different materials as two sides of the same fiber. This special morphology makes two different polymers available for property combinations, postspinning modifications, etc., in a single fiber, keeping the structure and properties of individual materials on two sides intact. The basic electrospinning apparatus remains the same, but special nozzles/spinnerets are used for making core–shell and side-by-side fibers (Figure 4.28). Nowadays these special spinnerets are commercially available; otherwise they can also be made in the own institute/university workshop.

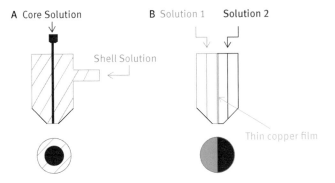

Figure 4.28: Schematic diagram of bicomponent electrospinning spinnerets: (A) coaxial and (B) sidebyside.

Materials

Polystyrene (PS 165 H BASF; M_w = 300,000 g/mol), poly(methyl methacrylate) (PMMA, Sigma-Aldrich, M_n = 120,000 g/mol), DMF (99.99%, Fisher Chemical, density: 0.948 g/cm³).

Bicomponent fiber preparation

30 wt% polymer–DMF solutions (PS and PMMA) were used for making core–shell (PS–core; PMMA–shell) and side-by-side fibers. 1.5 g each of PS and PMMA were dissolved in 3.5 g of DMF separately. The conductivity of the two solutions at 25 °C was as follows: PMMA–DMF 2.8 µS/cm and PS–DMF 0.9 µS/cm.

Electrospinning

The electrospinning parameters were as follows: applied voltage at the needle +19 kV, counter electrode –1 kV. Two injector pumps were used to control the flow rates of the two solutions (0.3 ml/h each). The collecting distance was 15 cm, and the fibers were collected on a baking paper. The RH was 28%.

Morphology

The average diameter of side-by-side and core–shell fibers was 4.2 ± 0.19 μm and 798 ± 71 nm, respectively. The morphology of the electrospun bicomponent fibers can be found in Figures 4.29 and 4.30. For getting a higher contrast between the two polymers, both hybrid fiber types were stained with RuO$_4$ (according to section 3.2.1.2) before SEM measurement. Therefore, the PS fraction appears brighter than the PMMA fraction in pictures.

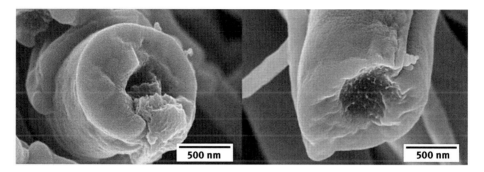

Figure 4.29: SEM images of electrospun coaxial PS/PMMA fibers. Core: PMMA; shell: PS. The fibers were stained with RuO$_4$ (according to section 3.2.1.2). Therefore, PS appears brighter than PMMA.

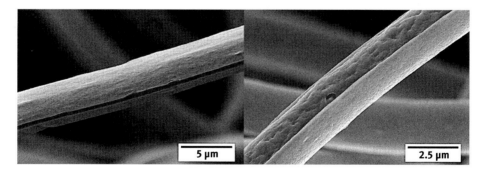

Figure 4.30: SEM images of electrospun side-by-side PS/PMMA fibers. The fibers were stained with RuO$_4$ (according to section 3.2.1.2).

4.8 Interesting to know

–   Poly(ethylene oxide) (PEO) is a water-soluble biocompatible polymer used for various biomedical applications and in cosmetic formulations. It is technically made by an anionic ring-opening polymerization as shown here:

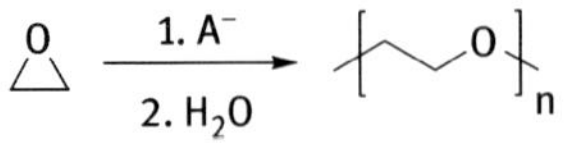

Ethylene oxide   Poly(ethylene oxide)

Scheme 4.6: Synthetic scheme for the formation of poly(ethylene oxide).

- Poly(ethylene glycol) (PEG) and PEO have the same chemical repeat unit structure $(-CH_2-CH_2-O-)$. The difference lies in the polymer synthesis procedure and the molar masses. PEO is a very high-molar-mass polymer (as high as 900,000 g/mol is possible), whereas PEG is available in low molar masses (20,000–30,000 g/mol). PEO is made by a ring-opening polymerization (ROP) of ethylene oxide, whereas PEG is made by a condensation polymerization, using the ethylene glycol monomer.
- Side-by-side nozzles are also used for making Janus particles (Figure 4.31) [8]. The apparatus used remains the same as for preparing Janus fibers, but the processing conditions, i.e., voltage and viscosity of the solution are different so that instead of fibers particles are formed.

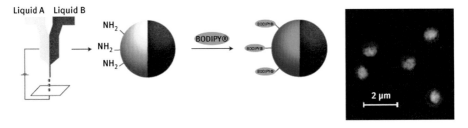

Figure 4.31: Formation of Janus particles using side-by-side nozzle and confocal microscopic images confirming side-by-side structure. Selective chemical modification on one side of the two-phasic nano/micro particles is also possible. [Reprinted with permission from *Nature Mater.* **2005**, *4*, 759. Copyright Nature Publishing Group (2005).]

- The Cox-Merz rule was discovered by W.P. Cox and E.H. Merz[21] and was empirically found to count for the steady state region (Newtonian region) of polymer melts and polymer solutions. It states equal values for the shear viscosity (η) plotted against the shear rate ($\dot{\gamma}$) and for the complex viscosity $|\eta^*|$ plottet against the angular frequency (ω) in the steady state region (see chapter 3.1.3). The rule can be an advantage, since zero shear viscosities can be estimated by oscillatory experiments, which don't have the difficulties of secondary flows or sample fractures. In the converse situation it is possible to evaluate the steady state complex viscosity via shear viscosity tests, if no access to oscillatory rheometer persists. The rule can also be an analytical benefit by comparing both values, which give insides in the microstructure of a sample since the rule only applies for unfilled (non-colloidal) systems. Further information can be found in the literature [9].

– Kirby-Bauer or agar diffusion test is generally used to evaluate the antibacterial activity and is also helpful in commenting about the leaching/non-leaching behavior of the antibacterial materials. The physical mixing of antibacterial material as an additive to a polymer might lead to simple leaching with time in contact with water/humidity. On the other hand, chemical immobilization of antibacterial moiety to a polymer chain provides non-leaching antibacterial material. In Kirby-Bauer test, the sample is placed on a nutrient agar plate inoculated with a definite amount of bacteria and incubated for 24 h at 37 °C. After incubation, the plates are visually investigated for formation of a zone of inhibition, which is an area without bacteria colony formation due to the release of antibacterial material. The zone of inhibition shows antibacterial activity of the material by leaching in the surroundings. In contrast, if bacterial colonies grow all over the agar plate including the area below the sample, implies no antibacterial activity.

References

[1] Ch. Hellmann, J. Belardi, R. Dersch, A. Greiner, J. H. Wendorff, S. Bahnmueller, *Polymer* **2009**, *50*, 1197.
[2] M. Richard-Lacroix, C. Pellerin, *Macromolecules* **2013**, *46*, 9473.
[3] L. Peng, M. Seuß, G. Lang, T. Scheibel, A. Fery, S. Agarwal, *Macromol. Mater. Eng.* **2015**, DOI: 10.1002/mame.201500217.
[4] K. Bubbel, Y. Zhang, Y. Assem, S. Agarwal, A. Greiner, *Macromolecules* **2013**, *46*, 7034.
[5] A. Gupta, S. Silver, *Nature Biotechnol.* **1998**, *16*, 888.
[6] S. H. Jeong, S. Y. Yeo, S. C. Yi, *J. Mater. Sci.* **2005**, *40*, 5407–5411.
[7] Q. L. Feng, J. Wu, G. Q. Chen, F. Z. Cui, T. N. Kim, J. O. Kim, *J. Biomed. Mater. Res.* **2000**, *52*, 662–668.
[8] K. H. Roh, D. C. Martin, J. Lahann, *Nature Mater.* **2005**, *4*, 759.
[9] T. S. R. Al-Hadithi, H. A. Barnes, K. Walters, *Colloid Polym. Sci.* **1992**, *270*, 40–46.

5 Selected applications of electrospun fibers and chemistry of corresponding polymers

A brief introduction to important terms, the chemistry of polymers often used for electrospinning and important application areas is given in this chapter.

5.1 Thermoplastics, elastomers and thermosets for electrospinning

5.1.1 General introduction

Thermoplastic polymers can be shaped and molded by heat above their respective glass transition temperatures or melting points. They can be amorphous or semicrystalline. Amorphous thermoplastic polymers are transparent for visible light which is due to their random disordered molecular structure. Thermal transitions of such polymers are characterized by a glass transition temperature (T_g), which marks the start of segmental motions and leads to polymer softening. Thermoplastics are hard and glassy materials at room temperature since the T_g is higher than room temperature [the respective T_g is dependent on many different factors such as molar mass, polymer architecture (linear, branched, etc.)]. Semicrystalline polymers show packing of some molecular segments in ordered regions, called crystallites. They show in addition to T_g a characteristic melting phase transition, represented by the melting temperature (T_m). In addition, semicrystalline polymers are opaque due to scattering of light. A semicrystalline polymer can also look transparent when the crystallite sizes are less than half the wave length of light. The crystallite size and the percentage of crystallinity of a semicrystalline polymer depend on factors such as polymer architecture, molar mass and polymer processing conditions. Have you ever wondered why poly(ethylene terephthalate) (PET) water bottles are transparent in spite of the fact that it is a semicrystalline polymer? Quench cooling (rapid cooling) of a semicrystalline polymer from melt leads to the formation of very small crystallites and therefore they appear as glassy transparent polymers and the same is true for PET bottles.

Furthermore, polymers with the same chemical structure can show different glass transition temperatures and crystallinity behaviors. For example, poly(methyl methacrylate) (PMMA) can have T_gs at ca. 60, 115 and 105 °C. Moreover, PMMA can be semicrystalline or completely amorphous. This is due to a difference in tacticities. Polymers based on vinyl monomers can show different tacticities and hence variations in physical properties like solubility, melting point, glass transition temperature and mechanical properties. Tacticity describes the spatial arrangement of substituents at chiral carbons, within a macromolecular chain (Figure 5.1).

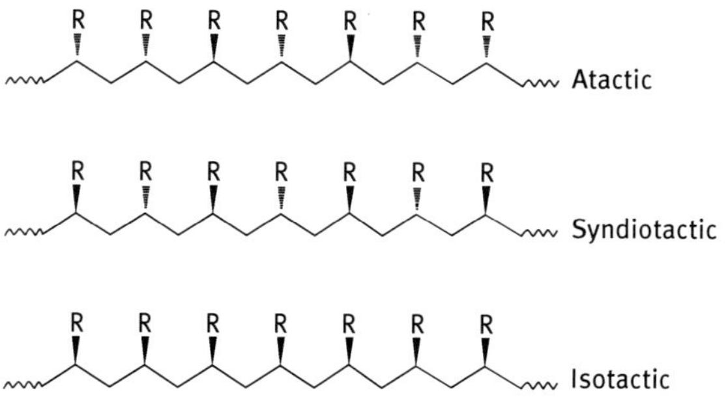

Figure 5.1: Tacticity of polymers. Atactic polymers are amorphous, whereas syndiotactic and isotactic polymers are semicrystalline.

The polymerization type and reaction conditions greatly influence the tacticity of polymers. Vinyl monomers undergo polymerization by addition of initiator units at the double bond. The most common initiators are radical, anionic, cationic or metal catalysts. Polymers made by a conventional radical polymerization [i.e., using radical initiators such as azobisisobutyronitrile (AIBN) and benzoylperoxide (BPO)] are mostly atactic and hence amorphous. An exception is low density polyethylene made by high temperature, high pressure radical polymerization, which is a semicrystalline polymer. Stereoregular, isotactic and syndiotactic polymers can be obtained by either metal-catalyzed or anionic polymerizations, under specific reaction conditions of temperature and solvent. The glass transition temperatures of PMMA are 60, 115 and 105 °C for isotactic, syndiotactic and atactic configurations, respectively. Isotactic PMMA is semicrystalline, with a melting point of around 160 °C. Moreover, it is opaque. The amorphous, atactic PMMA made by radical polymerization is an example for a classical transparent polymer, also called Plexiglas. ^{1}H, ^{13}C NMR and X-ray powder diffraction are some of the very easy and common methods used for the determination of polymer tacticity.

Linear thermoplastics made by condensation polymerization such as polyesters and polyamides are generally semicrystalline. In a condensation polymerization, di- or multifunctional monomers condense together with the release of small molecules such as water and ammonia depending on the starting materials used (Scheme 5.1). The use of branched difunctional monomers can provide amorphous condensation polymers depending on the length and type of side-chain branches. The condensation polymerization is also called a step-growth polymerization, since molar mass build-up takes place in steps. This is a contrast to radical, cationic, anionic, or metal-catalyzed vinyl polymerizations, which are classified as chain polymerization in which the build-up of molar mass is at the beginning of the polymerization.

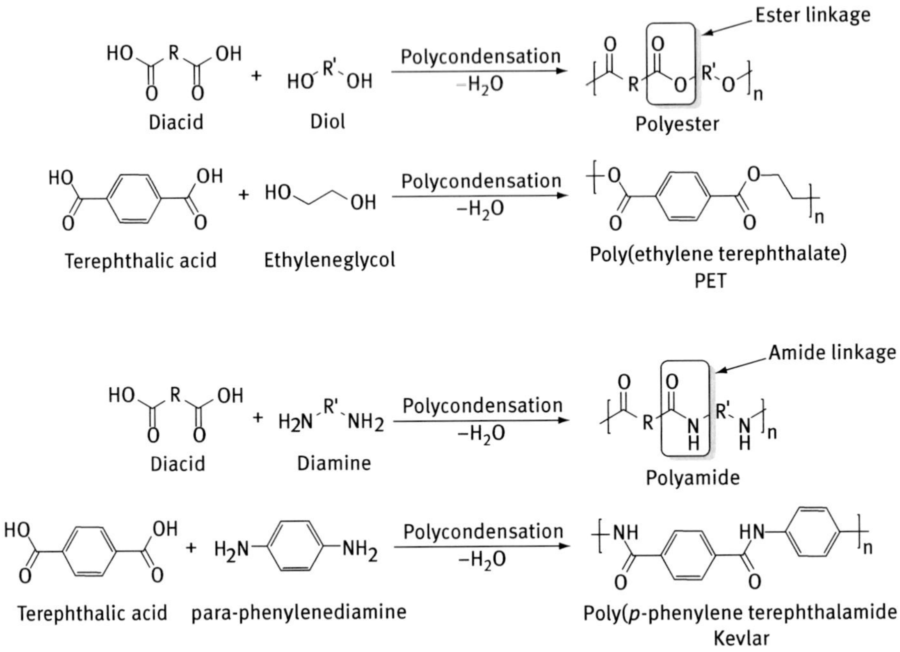

Scheme 5.1: Condensation polymerization for the formation of polyesters and poly(ar)amides. Poly (ethylene terephthalate) and poly(p-phenylene terephthalamide) are shown as representative examples.

Some of the condensation polymers like aliphatic polyesters, polyamides, and polycarbonates can also be made by ring-opening polymerization (ROP) method starting from the cyclic monomers as described in later sections with appropriate applications. ROP can produce very high molecular weights in contrast to condensation polymerizations. The molar masses obtained by condensation polymerization are generally limited to 30,000–40,000 g/mol. Chain extensions in a second step using appropriate chemistry are carried out to end up with high-molar-mass condensation polymers.

Some of the common amorphous and semicrystalline polymers are listed in Table 5.1. These polymers are soluble in various organic solvents, and many of them are very good candidates for making nanofibers by electrospinning for different applications. It is worth mentioning that although poly(vinyl alcohol) (PVOH) and polyacrylonitrile (PAN) are re-processable polymers and can be reused and reshaped like thermoplastics, they are not processed by conventional thermoplastic processing techniques such as injection molding, extrusion, compression molding and blow molding. The reason lies in the low degradation temperature (around 150 °C) of PVOH via elimination of water and the very high melting point (above 300 °C) of PAN. PAN degrades before melting making thermal processing impossible under standard conditions. Therefore, PVOH and PAN are processed mostly from solution not only for making fibers but otherwise also.

Table 5.1. Commonly used synthetic, amorphous and semicrystalline polymers for electrospinning.

Polymer	Repeat unit	Monomer(s)
Amorphous polymers		
Poly (methyl methacrylate)		
Poly (methyl acrylate)		
Polystyrene		
Poly (vinyl chloride)		
Poly (vinyl acetate)		
Poly (vinyl alcohol)		
Semicrystalline polymers		
Polycarbonate		
Polyacryloni-trile		
Poly(ethylene oxide)		
Nylon-6		
Nylon-6,6		

(continued)

Table 5.1. (continued)

Polymer	Repeat unit	Monomer(s)
Semicrystalline polymers		
PET		
Polycaprolactone		
Polylactide		

n stands for the number of repeat units, also called degree of polymerization. The total mass of the polymer divided by the mass of the repeat unit gives the degree of polymerization.

A further variation of step-growth polymerization is polyaddition reaction used for making commercially important types of polymers such as polyurethanes, epoxy-, phenol – formaldehyde, melamine – formaldehyde resins. Among these, polyurethanes are highly used for making fibers by electrospinning for applications in tissue engineering, drug release, and composites. They are made by addition of diisocyanates and diols and are soluble in various organic solvents. Therefore, they are good candidates for electrospinning. The availability of large numbers of diols and diisocyanates with different spacer groups provides polyurethanes with a variety in functionality, solubility, and properties.

Linear polyurethanes with blocks of two different types of polarities are called thermoplastic polyurethanes (TPU). The polar blocks can interact with each other via different macromolecular interactions, for example, via hydrogen bonding, leading to ordered crystalline structures, which are distributed in the amorphous matrix – made by the nonpolar block – in the form of a phase-separated structure. The phase-separated structures act as physical cross-linking points, which can be reversibly broken by heat and solvents. TPUs are processable like thermoplastics and show the elasticity and flexibility of elastomers. Therefore, they belong to a special class of polymers, called thermoplastic elastomers (TPEs).

In contrast, conventional elastomers are covalently (chemically) cross-linked polymers, which show a reversible stretchability and low Young's moduli. The term *rubber* is used interchangeably with elastomers. The covalent cross-linking of low T_g

polymers (T_g less than room temperature) provides conventional elastomers such as polybutadiene (BR), polyisoprene (IR, natural rubber), and polychloroprene (CR). The electrospinning of elastomers is mostly limited to TPUs as they are soluble in organic solvents. In contrast, spinning of conventional elastomers requires an *in situ* cross-linking as non-cross-linked elastomeric precursors are highly sticky, viscous liquids at room temperature. Tian et al. [1] have shown the formation of elastomeric fibrous composites of polybutadiene, isobutylene – isoprene rubber and silicone rubber with silver nanoparticles by electrospinning, followed by *in situ* UV cross-linking.

The general term used for cross-linked polymers is thermosets. Once they are set, they cannot be remoulded/reshaped. The polymer precursors are used for processing of thermosets and cross-linking is carried out either during or after processing. The difference between elastomers and thermosets lies in the degree of cross-linking, which is very high in thermosets. Although epoxy, melamine – formaldehyde, phenol – formaldehyde, unsaturated polyesters, and bismaleimide resins are classical examples of thermosets, polymers such as acrylates, polyurethanes, or dienes can also build up cross-linked networks. *Resin* is a term that is generally used for viscous low-molar-mass precursors, which are capable of undergoing cross-linking by polymerization. For example, the epoxy precursor made by reaction of bisphenol-A (BPA) and epichlorohydrin (ECH) is shown in Scheme 5.2. It undergoes further polymerizations using dinucleophiles such as diols, diamines, and diacids, leading to a cross-linked structure. The temperature at which cross-linking takes place depends on the cross-linking agent. Aliphatic amines can cross-link epoxy resins even at room temperature.

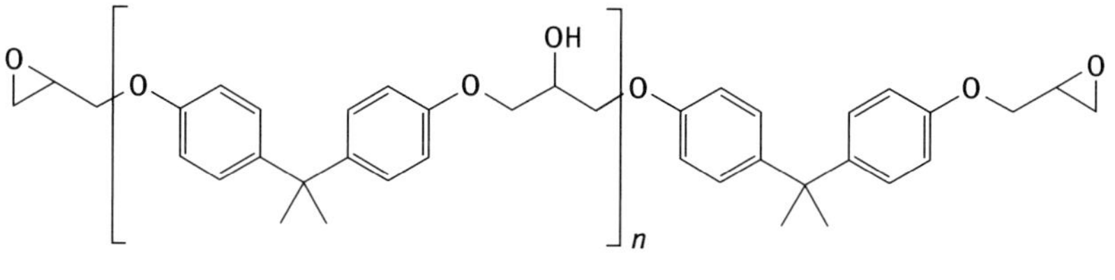

Scheme 5.2: Epoxy precursor for making cross-linked thermoset.

The cross-linking for making thermosets is carried out mostly by heat or light, although other methods are also possible. There are efforts in the literature to make cross-linked electrospun fibers for different applications via reactive electrospinning (Figure 5.2). This is a term used for cross-linking of electrospun fibers either during spinning or just after the collection of the fibers at the counter electrode. Conventional single-nozzle electrospinning setups can be applied for a reactive electrospinning with the provision of UV light exposure by putting UV lamp(s) at an appropriate position. For UV or thermal cross-linking, more than one polymerizable acrylate/methacrylate group at the polymer backbone or chain-ends is required. For instance, cross-linked polyethyleneimine and polycarbonate microfibers were prepared by a reactive electrospinning of methacrylated polyethyleneimine and polycarbonate in the presence of UV light. The polymers before electrospinning can be modified using the appropriate chemistry for the introduction of cross-linkable groups [2, 3].

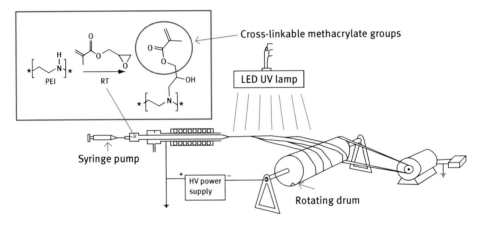

Figure 5.2: Reactive electrospinning: an UV cross-linkable polymer is spun with simultaneous UV exposure for cross-linking the fibers. [Reprinted by permission from *Biomacromolecules* **2010** Sep 13, *11*(9), 2283–2289. Copyright American Chemical Society (2010).]

Other options for reactive electrospinning could be the use of a two-component system, i.e., matrix polymer/precursor and a chemical cross-linking agent as the two components. For the handling of two-component cross-linking systems, special modified electrospinning nozzles are required. The two components – the matrix polymer that has to be cross-linked and a cross-linker – should come into contact with each other at the tip of the nozzle. It is worth to mention that the cross-linking reaction should be fast at room temperature, in order to provide efficient cross-linkings providing stable fibers just after spinning. The fibers can be further annealed by heat in the second step, for finalizing the cross-linking process, if required. An example was shown by Rafailovich et al. using a dual syringe system, supplying modified hyaluronic acid [thiolated HA derivative, 3,3′-dithiobis(propanoic dihydrazide)] and a cross-linker poly(ethylene glycol) diacrylate (PEGDA) (Figure 5.3) [4]. A Michael addition reaction of thiol at acrylate/methacrylate double bond was used for cross-linking.

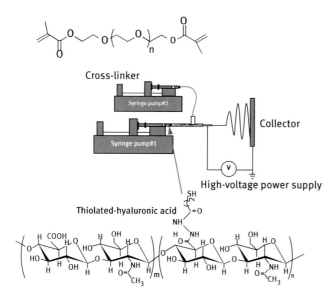

Figure 5.3: Reactive electrospinning: thiolated hyaluronic acid cross-linked with poly(ethylene glycol) diacrylate. [Reprinted by permission from *Macromol. Biosci.* **2006**, *6*, 811). Copyright Wiley-VCH Verlag GmbH & Co. KGaH Weinheim (2006).]

5.1.2 Polymers for selected applications of electrospun fibers

5.1.2.1 Filter applications

Nylons are highly used polymers for electrospinning in the case of filter applications. They belong to the class of polyamides. Nylon-6,6 is made by condensation polymerization of hexamethylene diamine and 1,6-hexanoic acid (adipic acid) (Scheme 5.3). In general, nylons are represented by two numbers; the first number shows the number of carbon atoms in the amine and the second number represents the total number of carbon atoms in the diacid (including two carbonyl carbons from the acid groups) used for making the corresponding nylon by condensation reaction of AA (diamine) and BB (diacid) monomers. For example, nylon-6,10 is made of hexamethylene diamine and sebacic acid (Scheme 5.3).

Nylon-6,12, nylon-6,14, nylon-10,12, nylon-6,13, nyon-6,15, nylon-10,11, and many other types are also known. High thermal stabilities and good mechanical properties make nylons a good candidate for electrospinning for filter applications. With an increasing number of methylene groups between diacids or diamines, i.e., in the higher nylons such as nylon-6,14, nylon-10,12, and nylon-10,14 hydrophobicity increases with a coincident decrease in stiffness, strength, melting temperature, and solubility in common organic solvents, such as formic acid and DMF. Therefore, higher nylons are less used for electrospinning.

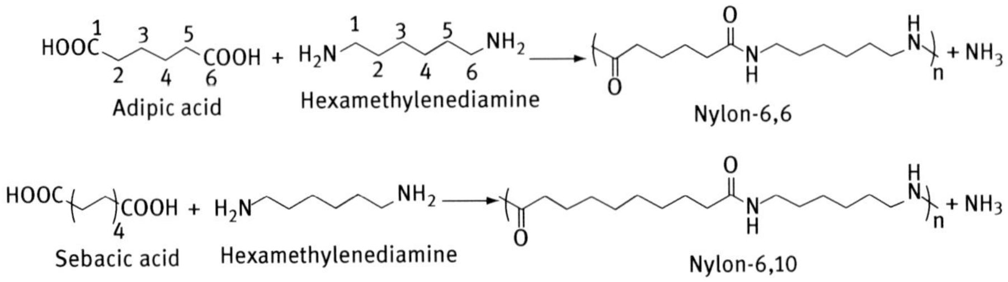

Scheme 5.3: Chemical structures and synthetic scheme for the formation of nylon-6,6 and nylon-6,10.

The nylons represented by only one number, such as nylon-6, are commercially made by ring-opening polymerization (ROP) of cyclic amides such as caprolactam (Scheme 5.4).

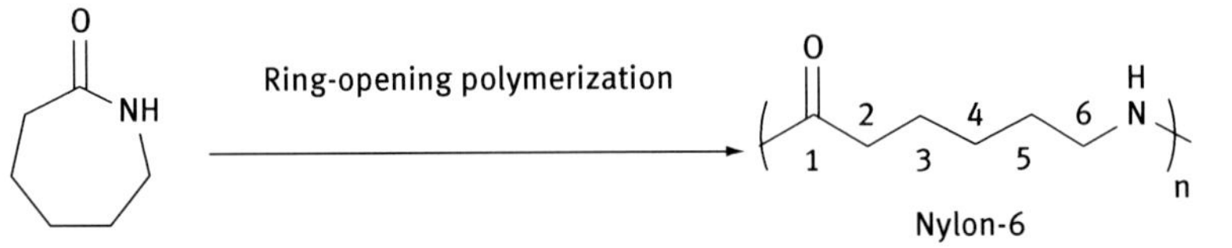

Scheme 5.4: Ring-opening polymerization of ε-caprolactam for the formation of nylon-6.

Nylon-6 and nylon-6,6 are the most common examples of polyamides used for electrospinning from DMF. One of the best established applications of electrospun fibers is in filtration such as HEPA (high efficiency particulate air) and ULPA (ultra low penetration air) filters. The small fiber diameters, enhanced surface areas, high porosities (defined as the ratio of pore volume to the total volume), and good interconnectivities of the pores are some of the advantages of electrospun fibrous nonwovens for filter applications, providing efficient filtrations with a reduced pressure drop and flux resistance. The size of the particles filtered and hence the filter efficiency is dependent on the pore size, which can be adjusted by changing fiber diameter. The mean pore size is generally 3–4 times the mean fiber diameter, whereas the maximum pore size could be 8–12 times the mean fiber diameter [5].

State-of-the-art air filters contain thick (micrometer) fiber networks with big pores. During air purification, the particles get trapped in the depth of the filter material, thereby increasing the pressure drop with time and cleaning cycles. The problem can be overcome by use of electrospun fibers with very small diameter. Coating a thin layer (5–10 μm thick) of electrospun nanofibers in very small amounts (0.5–1.0 g/m^2) onto conventional filter materials is generally done to improve the filter efficiency. A 0.5 g/m^2 coating of nylon-6,6 with an 100–200 nm fiber diameter showed more than 95% efficiency in filtering particles of size 0.16 to 0.94 μm without much decrease in the pressure drop (Figure 5.4) [6].

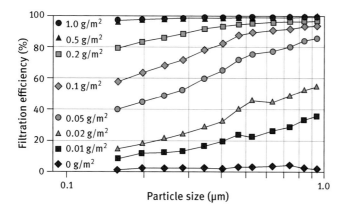

Figure 5.4: Filtration efficiency of a nylon-6,6 nanofiber-coated filter. A 0.5 g/m² coating showed filtration efficiencies of more than 95% for all particle sizes between 0.16 and 0.94 μm. A coating with 1 g/m² gave no additional advantage. [Reprinted by permission from *Polym. Eng. Sci.* **2008**, *48*, 1168–1176. Copyright Society of Plastic Engineers (2008).]

Coatings with electrospun nanofibers generate small pores, which even filter off very small particles also. The mechanism of filtration is entirely physical, by excluding particles of size larger than the average pore size at the surface, also called sieve effect. The same principle can also be applied for liquid micro-, ultra- and nano-filtrations [7]. For example, condensation and vinyl polymers such as polysulfones, nylons, polyvinylidene difluoride (PVDF) and polyacrylonitrile (PAN) were electrospun for liquid filtrations [7–9]. The choice of polymer material for making fibers could be critical in the case of liquid filtration as the polymer should be chemically stable and likewise mechanically strong. Moreover, in cases of hydrophobic polymers, the water permeability might be hampered for liquid filtration application. For increasing the hydrophilicity, an extra water-permeable top layer can also be applied [9].

Depending on the choice of polymer for electrospinning, additional properties in filters can be introduced. The use of Ag nanoparticles in polymer fibers can help in prevention of bacterial growth on filters as shown for PAN-Ag composite fibers [10]. Electrospinning offers an advantage over other fiber forming techniques in terms of making composite fibers in a simple way. The functional additives can simply be mixed/dispersed in the electrospinning formulations before spinning. Other modified electrospinning techniques such as coaxial and side-by-side spinning can also be used for making composite fibers as described in Chapter 2.

Poly(ethylene terephthalate) (PET), an aromatic polyester, is also a good candidate for filter applications due to the good chemical resistance and the high thermal and mechanical stability. It has been electrospun from a mixture of trifluoracetic acid (TFA) and dichloromethane (DCM) for fruit juice clarification [11]. Apart from polyamides made by condensation polymerization as described above, polyesters are another important class of thermoplastics. Aliphatic polyesters are widely electrospun to make scaffolds for tissue engineering. They are examples of biodegradable and biocompatible polymers and are prepared by ROP of the corresponding cyclic esters. They are described in detail in the next section.

5.1.2.2 Tissue engineering and drug-release applications

Aliphatic polyesters such as polylactide (PLA), polycaprolactone (PCL), and their copolymers with glycolide, poly(ethylene oxide), poly(hydroxybutyrate), and poly (hydroxyvalerate) copolymers, poly(vinyl alcohol), polyphosphazene and poly(ester urethanes) are some of the synthetic biodegradable polymers that have been electrospun for applications such as tissue engineering, controlled drug release and encapsulations of biomolecules, enzymes, viruses and bacteria [12]. Cellulose derivatives, silk fibroin, collagen, gelatin, hyaluronic acid, fibrin, fibrinogen, and elastin are examples of biobased biodegradable polymers, being used for similar applications. Poly(hydroxybutyrate), poly(hydroxyvalerate), and poly(hydroxybutyrate-co-valerate) (Scheme 5.5) are examples of the general class of biodegradable poly(hydroxyalkanoate) polyesters, which are produced by bacteria for energy storages [13]. They can also be manufactured synthetically.

Poly-(R)-3-hydroxybutyrate (P3HB)   Poly-3-hydroxyvalerate (P3HV)

Scheme 5.5: Chemical structures of poly(hydroxyalkanoate) derivatives.

According to the IUPAC definition, polymers which are susceptible to degradation by biological activity of microorganisms such as bacteria, fungi and algae are called biodegradable polymers. Biodegradation can happen in soil, water, animals and human beings. It is a property that depends on the chemical structure and not the origin. Here, the origin means petrochemical- or biobased. Both petrobased and biobased polymers can be biodegradable. The biodegradation of a polymer is accompanied by a lowering of its molar mass with a decrease in mechanical properties and mass loss. The first step of biodegradation takes place at the outside of the microorganism body. Oligomers with functional chain-ends, formed by polymer cleavage at hydrolytically or enzymatically sensitive functional groups in the polymer backbone, in the first step undergo metabolism in microorganism with an ultimate formation to carbon dioxide and water or methane. Polymer backbones with functional groups such as esters, carbonates, anhydrides and phosphates can provide biodegradability depending on molar mass, crystallinity, hydrophilicity – hydrophobicity, porosity, etc. Vinyl polymers having a C–C backbone are in general not biodegradable. Exceptions are poly(vinyl alcohol) and cis-polyisoprene. Details about biodegradation and biodegradable polymers can be seen in the review articles [14,15].

Ring-opening polymerization is widely used for making very well-known bio-degradable polymers such as PLA, PCL, their copolymers with polyglycolide and polycarbonates on a technical scale using metal catalysts. Tinoctanoate [tin(II) 2-ethylhexanoate; Sn(oct)$_2$] is one of the metal catalysts employed for this purpose. Other catalysts based on lanthanides, transition metals, etc. can also be utilized. If a polymer has to be availed for biomedical applications, it is important to use FDA (food and drug administration) approved catalysts as it is not possible to remove 100% of the catalyst after a reaction from the reaction product. The traces of catalyst remaining in the reaction product could be critical for biomedical applications.

Poly(ester urethanes) made by chain extensions of diol chain-ends substituted aliphatic polyesters by diisocynates, are also highly utilized biodegradable biomaterials, not only as scaffolds but also as implants, carrier for drug encapsulation, etc. A synthetic scheme for their preparation is provided in Scheme 5.6.

(DegraPol®) is one of the poly(ester urethane)s based on block copolymers of poly((R)-3-hydroxybutyric acid)-diol and poly(ε-caprolactone-co-glycolide)-diol, linked with a diisocyanate for making fibrous scaffolds in skeletal muscle tissue engineering [16]. Poly(ester urethanes) of varied properties can be easily generated by changes in either polyester or isocyanate structures. Poly[(L-lactide-co-ε-caprolactone)-co-(L-lysine ethyl ester diisocyanate)-block-oligo(ethylene glycol)-urethane] electrospun scaffolds were shown as a promising candidate for soft tissue engineering by growing adipose-derived stem cells (ASCs) [17].

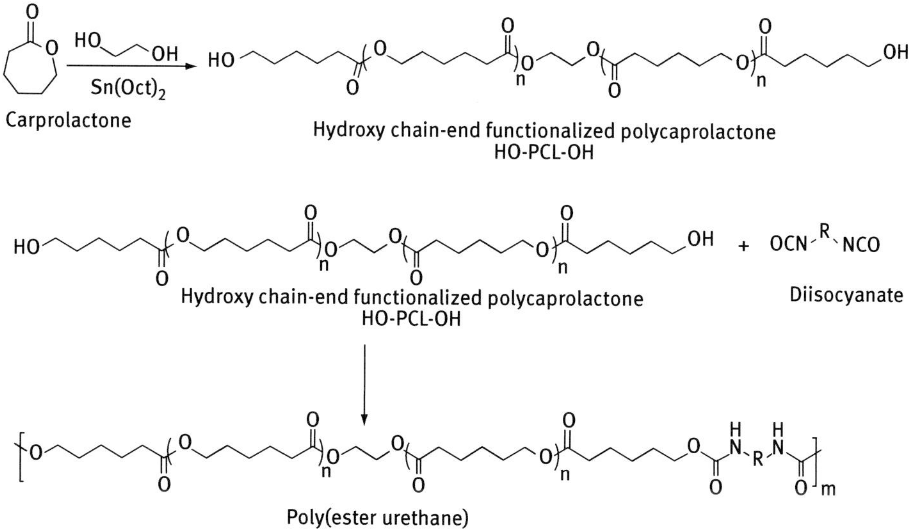

Scheme 5.6: Synthesis of poly(ester urethane).

Tissue engineering or regenerative medicine deals with the regeneration of new extra cellular matrix (ECM) that has been destroyed by disease, injury or congenital defects [18]. It requires biodegradable and biocompatible scaffolds, having high surface areas and 3D porous structures as support for cells. The use of electrospun biodegradable and biocompatible polymers for tissue engineering (targeted tissues: lung, heart blood vessels, skin, bone, nerve, etc.) is of significant interest as the resulting fibrous structures with high porosities can mimic the natural ECM. The proliferation and growth of many different types of cell-lines such as endothelial cells, coronary artery smooth muscle cells, schwann cells, glia, vero cell migration, human neural cells, osteoblasts, neural stem cells, and cardiomyocytes has been studied. The type of the regenerated tissue puts demands on the choice of material in terms of mechanical properties, degradation rate and biocompatibility. Electrospinning provides lots of material opportunities, as spinning conditions are already optimized for many different biodegradable polymers and various techniques for surface modifications of fibers are also well established.

Fiber morphology, orientation, and pattern can influence to a large extent growth, viability, differentiation potential, and penetration of cells in the bulk of the scaffolds. In one of the studies, the effect of fiber orientation of PCL on the extent of microglial and astrocytic response was studied by implanting electrospun scaffolds into the caudate putamen of an adult rat brain. There was no influence of the fiber orientation on the inflammatory response, but it eminently affected the neurite infiltration. Randomly oriented fibers enabled neurite infiltration and growth in contrast to aligned fibers [19]. In another study, the direction of neural stem cells elongation and its neurite outgrowth was parallel to the direction of aligned PLLA fibers [20]. Furthermore, human coronary artery smooth muscle cells (SMCs) preferably migrated along the axis of the aligned poly(L-lactide-co-ϵ-caprolactone) nanofibers [21]. This shows that every new cell-type and a new scaffold material should be individually tested for cell growth and proliferation. It would be too early to make any universal conclusions based on the literature data.

One of the problems which is often encountered in the use of electrospun fibers as scaffolds is the proliferation and growth of the cells only at the scaffold surface, lacking proper in-depth penetration due to the very small size of the pores. An optimum pore size is required for an in-depth cell penetration and attachment, depending on the cell types [22]. For example, a mean pore size of 100–150 µm facilitates cell attachments for bone tissue engineering. On the other hand, too big pores limit the specific surface area for cell attachments. The use of microfibers provides bigger pores leading to better cell penetrations, attachments and growths as shown by Hutmacher et al. for polycaprolactone-type I bovine collagen (µmPCL/Col) and µmPCL/Col-Heprasil fibrous scaffolds with fiber diameters of 1–2 µm (Figure 5.5) [23]. This shows that the use of nanofibers is not always advantageous and varies from application to application.

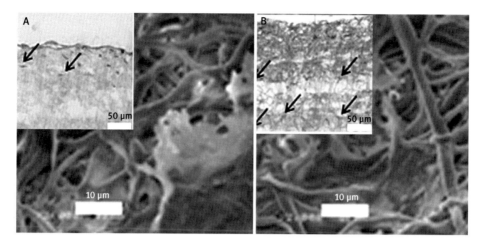

Figure 5.5: SEM pictures of (A) polycaprolactone-type I bovine collagen (μmPCL/Col), and (B) the μmPCL/Col-Hep (simultaneous electrospinning and electrospraying of a mixture of polycaprolactone-type I bovine collagen and Heprasil, respectively) scaffolds at 10 days post seeding human fetal osteoblasts showing cellular infiltration. The fiber diameter was 1–2 μm. The inset presents cryosectioning of the same samples with a hematoxylin and eosin (H&E) staining. The cells penetrated half the thickness of the fibrous scaffold in μmPCL/Col whereas cells penetrated the entire thickness in μmPCL/Col-Hep. [Reprinted by permission from *Biomacromolecules* **2008**, *9*, 2097–2103. Copyright American Chemical Society (2008).]

The conventional methods for making 3D tissue engineering scaffolds are by computer-aided designs, called rapid prototyping (RP) such as stereolithography, 3D printing and plotting, laser sintering or fused deposition modelling [24]. The pore sizes of scaffolds made by RP are relatively large in comparison to the cell size. This reveals problems in cell attachments. Rapid prototyping can be combined with electrospinning to get an optimized pore size that will provide benefit in cell in-penetrations and attachments. The macrostructure obtained from rapid prototyping fabrication techniques [3D fiber deposition (3DF)] was combined with microfibers by electrospinning by Moroni et al. to make scaffolds for cartilage tissue engineering using polyether-esters [block copolymers of poly(ethylene oxide)-terephthalate (PEOT) and polybutylene terephthalate (PBT)] (Figure 5.6). The combined structure showed higher proliferations, attachments and differentiations of chondrocytes, providing higher amounts of extracellular matrix in comparison to scaffolds with macrostructure produced only by rapid prototyping [25].

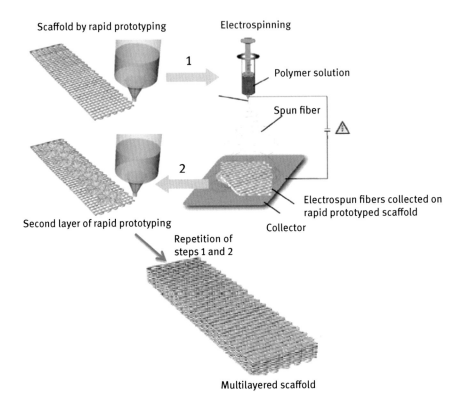

Scaffold by rapid prototyping

Electrospinning

Polymer solution

Spun fiber

Electrospun fibers collected on rapid prototyped scaffold

Second layer of rapid prototyping

Collector

Repetition of steps 1 and 2

Multilayered scaffold

Figure 5.6: Schematic drawing for the combination of rapid prototyping (3DF) and electrospinning for making multilayered scaffolds. [Reprinted by permission from *Adv. Funct. Mater.* **2008**, *18*, 53–60. Copyright Wiley-VCH Verlag GmbH & Co. KGaH Weinheim (2008).]

Cells can also be sandwiched between fibrous structures made by electrospinning, using a layer-by-layer (LBL) procedure. In the LBL method, cells are sprayed onto electrospun fibers. The number of fibrous layers, cell types and cell amounts can be easily controlled to provide 3D multilayer cell encapsulated fibrous scaffolds with uniform cell growths [26, 27].

Recently, a novel concept for making 3D scaffolds with big pores was presented by use of short electrospun fibers, after processing them from dispersions. A freeze-drying of the short-fiber dispersion provided a dual pore structure as observed by SEM: very large pores (300–430 µm diameter) and small pores (10–30 µm) within interconnected short fibers of aspect ratio 120–150 (Figure 5.7) [28]. The highly stable 3D spongy structure showed a promising future for encapsulations and proliferations of cells besides many other fields of application such as oil absorption and catalysis.

Figure 5.7: (A) Photograph of an ultralight sponge made from dispersion and freeze-drying of short electrospun fibers. (B) SEM image of the same sponge showing a dual pore structure. (C) Three-dimensional confocal microscopic image of Jurkat cells inside of the sponge (x, y, z = 450, 450, 300 μm). [Reprinted by permission from *Adv. Funct. Mater.* **2015**, *25*, 2850–2856. Copyright Wiley-VCH Verlag GmbH & Co. KGaH Weinheim (2015).]

Another application area where the use of electrospun fibers is very promising is for drug release. Electrospinning offers various ways for the encapsulation of drugs. The simplest way would be to simply mix the desired drug in the spinning solution. In many cases a burst-release of the drug was observed which might be due to the surface absorbed drug or by dissolution of fiber segments with time. Bicomponent electrospinning methods such as coaxial spinning, combined with a proper polymer, allow the encapsulation of drugs in the core of core-shell fibers, reducing the burst-release effect. This also controls release kinetics. The large number of electrospinnable polymers available with variation in chemical structure, hydrophilicity – hydrophobicity, etc., provides lots of opportunities for the tuning of the drug-release profile. Even the use of another bicomponent electrospinning technique, i.e., side-by-side electrospinning offers the advantage of adjusting the drug-release profile either for two independent drugs (two drugs are present in two different sides of a side-by-side fiber) or the same drug with different release kinetics (choose polymers of different degradability, solubility, hydrophilicity – hydrophobicity, etc., on two sides of the same fiber). The drug loaded fibrous mats can be used for wound healing. For such application, the porosity of fibrous mats offers additional advantages as it allows the permeation of oxygen, which is required for tissue regeneration during the wound healing process but hinders the penetration of bacteria [29]. The use of responsive polymers can provide external stimuli (temperature, pH, etc.) triggered controlled drug release [30, 31].

5.1.2.3 Gas-sensing, fuel-cell and battery applications

Electrospun fibers supply more sensitive gas sensors in comparison to the corresponding films for a large number of gases such as NH_3, water vapors, H_2S, O_2, CO, NO_2 and many organic vapors. Polar polymers such as poly(acrylic acid), poly(vinyl alcohol), poly(ethylene imine), conducting polymers and semiconducting metal oxides are used for making fibrous gas sensors with a fast response and recovery and excellent sensitivity utilizing different sensing methods and principles. For example, poly(acrylic acid) and poly(ethyleneimine)/poly(vinyl alcohol) fibers coated on a quartz crystal microbalance (QCM) were tried as acoustic wave sensors for NH_3 and H_2S and could detect concentrations as low as 150 ppb and 500 ppb, respectively [32, 33].

The change in resistance of conducting polymers can also be employed for monitoring the presence and amount of a particular gas. Conducting polymers such as poly(diphenylamine), polyaniline and poly(o-toluidine) with or without dopants can be used for this purpose. The most common dopants are: camphorsulfonic acid and multiwalled carbon nanotubes among many others. Inorganic fibers based on semiconducting metal oxides such as TiO_2, SnO_2 and ZnO can also be applied as resistive gas sensors. The inorganic fibers are made by electrospinning of a polymer solution [e.g., poly(vinyl pyrrolidone), PEO, PVOH, poly(vinyl acetate)] containing the corresponding metal oxides or precursor salts. The resulting fibers are annealed and sintered at high temperatures for getting metal-oxide fibers [34]. Metallic and bimetallic nanowires are also made in similar ways and show an edge over state-of-the-art carriers in electrocatalysis for fuel-cell applications. Conducting polymers such as polyaniline can also be used as carrier for immobilizing electrocatalysts.

Electrospun membranes of an appropriate polymer with high porosity, mechanical stability, thermal stability and flame resistance are suitable as electrode separator and carrier for liquid electrolytes for utilization in a Li-ion battery. The two electrodes are separated by a macroporous polyolefin membrane: mainly polyethylene and polypropylene to prevent short-circuits in the state-of-the-art Li-ion batteries. Low porosities, poor wettabilities of polyolefins towards liquid electrolytes and low thermal stabilities lead to a demand for new membranes for Li-ion batteries. Electrospinning combined with appropriate polymers can provide a solution to these problems. One class of suitable polymer candidates for this application is the class of polyimides (PIs). Polyimides are insoluble and infusible polymers. The melting/softening points are high and lie above the decomposition temperature, which is generally more than 400 °C. The PI fibers are made in two steps, starting from the soluble precursor poly (amic acid) (PAA) as described in Chapter 4. PAA fibers are heated with a programmed heating profile and rate to give PI fibers. This step is called imidization.

The reactions for the formation of PI are shown in Scheme 5.7. Poly(vinylidenefluoride-hexafluoropropylene) (PVDF-HFP)/poly(vinylchloride) or polyacrylonitrile are also suitable for the use as separation membranes in Li-ion batteries.

The high surface area of electrospun membranes is also useful in increasing the efficiency of microbial fuel cells (MFC). A MFC makes use of microorganisms such as *Geobacter* or *Shewanella* by oxidations of organic matters for the production of electric power. The very high anode current density ($30 A/m^2$) was achievable from electrospun carbon fiber electrodes [35].

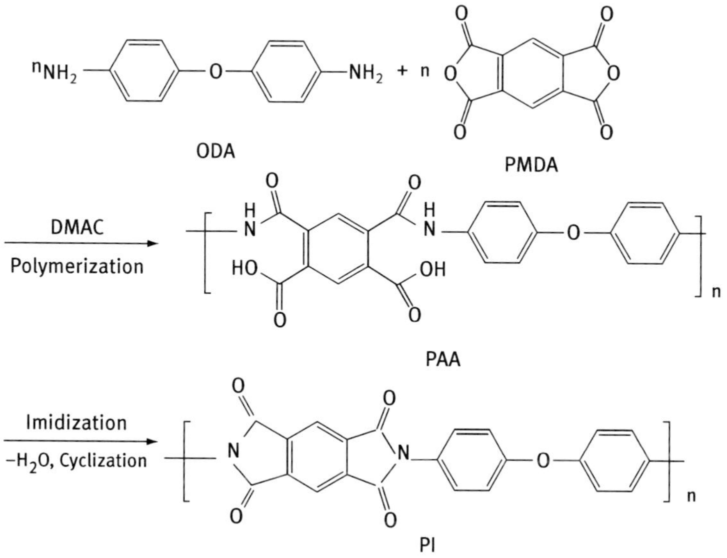

Scheme 5.7: Synthetic scheme for the formation of a polyimide. The representative example shows the reaction of pyromellitic dianhydride (PMDA) and 4,4'-oxydianiline (ODA) in dimethylacetamide (DMAc) for the formation of the precursor poly(amic acid) (PAA) in the first step, followed by an imidization for the formation of polyimide (PI) by loss of water molecules.

5.1.2.4   Polymeric electrospun fibers as reinforcing materials

As we have learnt till now, electrospinning can provide diverse polymeric fibers of different diameters and morphologies in a simple way. Unfortunately, the electrospun fibers in general lack molecular orientation and chain extension and therefore show weak mechanical properties with low tensile strength and modulus. Sometimes these values are even less than the corresponding bulk material and the microfibers made by state-of-the-art methods. Mechanical properties of elecrospun nonwovens can be improved, for example by smaller fiber diameters, molecular orientation, and alignment of fibers on a rotating collector, postdrawing or collection on a metal frame. The special case is the first strongest electrospun fiber mat, a polyimide [poly(p-phenylene biphenyltetracarboximide)] mat imidized at 430 °C. The rigid polymer structure led to orientation at high imidization temperature providing the highest strength and modulus of 650 MPa and 15 GPa, respectively. The corresponding single fibers showed strength and modulus as 1.7 GPa and 76 GPa, respectively. Copolymerization can further help in adjusting tensile strength, modulus and elongation at break for polyimide fibers [36]. Such high strength and modulus fibers can be used for reinforcing other thermoplasts and thermosets by making polymeric composites. Polymer composite is a general term used for a combination of a polymer matrix together with an additive which can be reinforcing fibers, functional additives such as antibacterial materials, metal nanoparticles and inorganic fillers. Fiber-reinforced composites are generally prepared to enhance the mechanical properties of the matrix polymer. Nanofiber-reinforced composites are also beneficial due to optical transparency of the resulting composites. Some of the important factors for getting fiber-reinforced polymer composites with improved properties are the use of mechanically strong reinforcing fibers, large aspect ratio of fibers, good wettability and interface between the fibers and matrix, and homogeneous distribution of fibers in matrix without aggregation. In order to improve matrix-fiber interface adhesion fibers with appropriate functional groups capable of chemically interacting with the matrix material can be used. Here, electrospinning is advantageous as many different functional polymers can easily be converted to fibers. For example, the fibers made up of a copolymer of styrene and glycidyl methacrylate will interact chemically with the epoxy matrix in presence of a diamine.

The fiber-reinforced polymer composites can be prepared in different ways. The easiest method of making electrospun fiber-reinforced composites is by dip-coating, i.e., the reinforcing fiber mat is dipped in a solution of the material to be reinforced and then dried. Many of such plies can be pressed together at an appropriate temperature and pressure using a compression molding machine. Composites reinforced by electrospun fibers can also be prepared by liquid filtration method, i.e., the fiber mat is sandwiched between two ordinary cellulose filter papers and matrix resin in appropriate solvent is sucked through the fiber membrane using a very small vacuum [37].

In most of the cases, electrospun fibers in the form of a nonwoven membrane are used for reinforcement purposes. Composite preparation by simple mixing of electrospun fibers either in solution/dispersion or by melt extrusion is not possible due to aggregation of long and continuous entangled fibers. Another option of making composites is by chopping electrospun fibers to smaller lengths for mixing either in polymer dispersion or in melt [38].

By the use of a mechanical stirrer with sharp blades, large quantities of electrospun fibers can be chopped to smaller lengths. Other methods of making short electrospun fibers are: photolithography (using photomask during cross-linking of fibers and dissolving the uncross-linked part), laser and ultrasound cutting, direct formation during electrospinning by choosing specific spinning conditions and materials, etc. [39]. The use of electrospun fibers in making high performance composites is handicapped by their low mechanical properties for the reasons described above. Future research efforts are required for getting strong electrospun fibers to make them available for high demanding applications.

5.1.2.5 Electrospun fibers as support for catalysts

The high porosity and surface area of electrospun fibers make them suitable as carrier for catalytic purposes. An additional advantage is the ease of catalyst immobilization and better mass – transport in fibrous supports. Both inorganic and polymeric fibers can be used depending on the reaction to be catalyzed. Inorganic fibers such as CeO_2, Cu/CeO_2, and palladium-copper-ceriumoxide ($Pd/Cu/CeO_2$), useful for methane oxidation, can easily be made by electrospinning of the precursor salts, followed by calcination [40, 41]. Noble metal nanoparticles and metal oxides such as Pt, Pd and TiO_2 can be simply immobilized on electrospun polymeric and inorganic fibers for various catalytic reactions [42]. The catalysts can also be simply mixed with electrospinning solution for immobilization. The challenge in designing such immobilized catalysts is to avoid the leaching into the reaction medium. Therefore, core-shell morphologies of fibers with catalyst as core material could be useful. The shell can be made directly by coaxial electrospinning or in a second step by coating the fibers with the immobilized catalyst by dip-coating, spray-coating, chemical vapor deposition, etc. In one of these studies, the leaching of scandiumtriflate catalyst from polystyrene fibers was controlled by coating of the fibers with poly(p-xylylene) (PPX) using a CVD technique [43]. PPX formation by CVD is a gas-phase polymerization technique, involving vapor phase pyrolysis of the starting monomers (paracyclophanes) at high temperature (600–700 °C) followed by condensation of corresponding radicals at room temperature on the substrate (Scheme 5.8) [44]. The CVD process requires no catalyst or solvent and the polymerization takes place at room temperature.

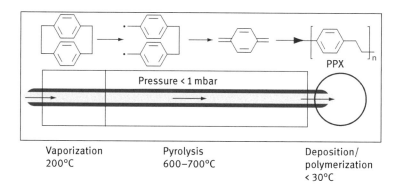

Scheme 5.8: Chemical vapor deposition (CVD) process for the formation of poly(p-xylylene) (PPX).

As mentioned in Chapter 2, the coating of electrospun fibers by PPX followed by thermal degradation/dissolution of the electrospun fibers provides hollow nano/micro-tubes (TUFT process). The method can be used for the immobilization of catalysts inside of the hollow tubes. These hollow tubes can be used as nano/microreactor. For example, gold nanoparticles could be immobilized in hollow PPX tubes by electrospinning of gold nanoparticles together with a template polymer for various catalytic reactions such as a hydrolytic oxidation of dimethylphenylsilane and an alcoholysis of dimethylphenylsilane with n-butanol. Any polymer with low degradation temperatures or solubilities – preferably in water or low boiling solvents – can be used as a template polymer. The gold immobilized template polymer fibers are coated in a second step with PPX by CVD. The degradation or dissolution of the template polymer will leave behind gold nanoparticles in PPX hollow tubes (Figure 5.8). The resulting gold-PPX nonwoven nanoreactors can be handled as tea-bag-like catalysts and can be easily removed from the reaction mixture and are reusable many times [45].

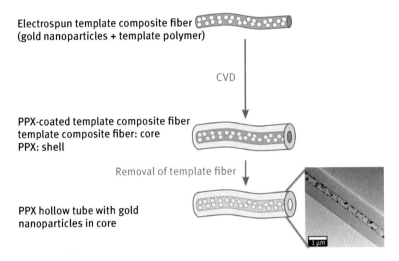

Figure 5.8: Schematic for the formation of gold nanoparticle – immobilized poly(p-xylylene) nanotubes. [Reprinted by permission from *Angew. Chem. Int. Ed.* **2014**, *53*, 4972. Copyright Wiley-VCH Verlag GmbH & Co. KGaH Weinheim (2014).]

5.1.2.6 Emulsions and dispersions for agricultural and biomedical applications

For agricultural applications, electrospun fibers can be used as carrier of pesticides, fertilizers, etc. for slow release applications in agricultural fields. The active components can be either directly mixed in electrospinning solution to get active monolith fibers or can also be encapsulated in the core of a core-shell fiber using coaxial nozzles for controlling the release rate. Another method of controlling the release rate is to encapsulate active components in polymeric microparticles and immobilize them on fibrous support using conventional electrospinning. The same is also true for immobilization and release of drugs for pharmaceutical applications. The carrier electrospun fibers are used either in the form of prefabricated nonwoven or directly spun at the site of use. Handheld electrospinning machines and special spinning arrangements on modified tractors are useful tools (Figure 5.9) for direct spinning in agricultural fields.

Figure 5.9: Handheld electrospinning machine with electrospun fibers on a grape vine plant (*left*) and modified tractor with spinning nozzles (*right*) are promising tools for direct spinning in agricultural fields.

Although solvent evaporation during fiber formation is very fast, the traces of organic solvents left in electrospun fibers could still be toxic when it comes to prefabrication of electrospun fibers using formulations based on organic solvents. In case of direct spinning on plants in agricultural fields (and also on human beings for wound healing applications), the use of organic solvents is not desirable. Figure 5.10 shows the effect of ethanol present in a spraying formulation made up of biodegradable and biocompatible poly(L-lactide) (PLA) electrospun short-fiber dispersion in water – ethanol mixture on the growth of *Arabidopsis thaliana* plants after three weeks exposure. The long and continuous fibers of PLA made by electrospinning are cut and dispersed in a water – ethanol mixture for this experiment. 40 mg PLA short fibers with a diameter of around 300 nm and a length between 10–200 micrometers were dispersed in a mixture of 3.5 ml water and 2.5 ml ethanol and further diluted to appropriate concentration for tests. The biomass production was reduced in comparison to the control due to the presence of ethanol (Table 5.2).

Water based electrospinning formulations are highly desirable and should be used for biomedical applications, wound healing, tissue engineering, agricultural areas, etc. Simple water soluble polymers cannot be used for many of these applications as the electrospun nonwoven should be stable in contact with water or physiological solution. The fibers obtained from water soluble polymers can be stabilized against dissolution in water by cross-linking as described in Section 5.2. The choice of cross-linking agents and method used for cross-linking depends on the application. Many applications cannot tolerate harsh cross-linking conditions such as UV light, the use of very high temperatures and special chemicals for cross-linking. Another way of spinning nonwater-soluble polymers from water is by the use of polymeric suspensions/dispersions.

Table 5.2. Biomass of *Arabidopsis thaliana* after 3-week exposure of 50 mg/L of tested compounds in comparison with control untreated plants.[*]

Sample	Biomass production (%) (in comparison with control)	Standard deviation SD
Ethanol	64	28
PLA+ethanol	68	34
PLA+surfactant[a]	12	5
Surfactant	21	10
MPEG-PHA[b]	88	46

[*]The data were collected by Dr. Premysl Landa, Laboratory of Plant Biotechnologies, Institute of Experimental Botany AS CR, Czech Republic.
[a] Cetyltrimethylammnium bromide was used as surfactant.
[b] MPEG-PHA = methoxy-terminated poly(ethylene glycol) block copoylmer with poly(hexamethylene adipate).

Figure 5.10: *Arabidopsis thaliana* plants after 3-week exposure to tested compounds at the concentration of 50 mg/mL. From left to right: control; PLA fiber dispersion in ethanol – water mixture; ethanol. The data is provided by Dr. Premysl Landa, Laboratory of Plant Biotechnologies, Institute of Experimental Botany AS CR, Czech Republic.

A suspension is a mixture of fine solid particles in a dispersing medium, for example, water. In this case a suspension is also called dispersion, since a dispersion is only the description of systems with multiple phases. Polymeric suspensions, in which small polymer particles are dispersed in water, are mainly made by special polymerization techniques, such as suspension and emulsion (micro- and miniemulsion) polymerizations, using radical initiators. The monomers are dispersed in water and polymerized using monomer soluble radical initiators in the case of a suspension polymerization. To hinder the coalescence of polymer particles, small amounts of a stabilizer are needed. Common stabilizers for suspension polymerization are, for example poly(vinyl alcohol), poly(vinyl pyrrolidone), and cellulose ether. In emulsion polymerizations micelles are formed by emusifiers which are aggregates of about 50–100 molecules. The monomers are initiated by the water soluble radical initiator in the aqueous phase and diffuses into the micelles with increase in chain length. Further monomers diffuses into the micelles and gets polymerized in the form of particles (100–200 nm) suspended in water. The dispersions made by suspension/emulsion polymerization are also called primary dispersion or latex. A large variety of primary dispersions such as polystyrene and its copolymers, acrylates/methacrylates, poly(vinyl chloride), poly(vinyl acetate), fluorinated polymers, styrene-butadiene, or acrylonitrile-styrene-butadiene are commercially available and can also be made easily in research laboratories. Another method for the production of aqueous polymeric dispersions is to disperse a readymade water-insoluble polymer in water. The resulting dispersion is named as a secondary dispersion. Several methods are available for the preparation of secondary dispersions. The emulsification of a polymer solution in an organic solvent in water containing surfactants is one of the simplest methods. The surfactants are amphiphilic molecules with hydrophilic heads and hydrophobic tails and prevent aggregation and agglomeration of polymer particles. They are generally based on fatty acid hydrophobic chains, amino acids, and poly(ethylene oxide). Polymer dispersions are formed once the organic solvent has been evaporated. The size of the particles in a dispersion can be controlled by using homogenizers and high speed stirrers. The dispersion obtained by this method has a generally very low solid content which can be increased to as high as 40–50% in a second step, i.e., by dialysis. Nanoprecipitation is another technique for making secondary dispersions. Thereby a polymer is precipitated in the form of tiny particles in water from a water miscible organic solvent. Secondary dispersions can also be made without using organic solvents. A thermoplastic polymer can be heated above its melting point and the melt can be emulsified in water which contains a stabilizer.

Polymeric dispersions are highly interesting for making fibers by electrospinning due to several reasons. First of all, aqueous dispersions provide safe, water based electrospinning formulations and allow formation of fibers without using toxic and flammable organic solvents such as DMF, HFIP, THF, and $CHCl_3$. This is also highly relevant from the point of view of safety issues, i.e., avoiding explosions and exposure to large amounts of organic solvents during spinning in both lab-scale small and up-scaled machines. We should not forget that the spinnable concentration in most of the conventional electrospinning formulations is most of the time less than 10–15 wt%. This implies that 85–90% of these formulations is solvent. The primary and secondary dispersions can be made with high solid contents of about 40–50 wt%. The use of high solid content dispersions for electrospinning also increases the productivity. The choice and concentration of surfactant for the stabilization of dispersions could be critical for agricultural and biomedical applications. Some surfactants can provide unwanted toxicity. A large number of commercially available and specially designed surfactants is known. They can be neutral, anionic or cationic, depending on the polar head. It is important to use biocompatible, nontoxic surfactants if emulsions are intended for biomedical and agricultural applications. Some of the biocompatible surfactants are: Tween 20 and Tween 80, Span

80 (nonionic) and pluoronics (PEO-PPO copolymers). Toxicity of surfactants for plants is proved in an experiment by a significant decrease in biomass production (Table 5.2 and Figure 5.11) by taking a water based dispersion of a biocompatible polymer PLA fibers stabilized by cetyl trimethyl ammonium bromide.

Figure 5.11: *Arabidopsis thaliana* plants after 3-week exposure of tested compounds at the concentration of 50 mg/mL. From left to right: PLA fiber dispersion with surfactant (cetyl trimethyl ammonium bromide); MPEG-PHA; control. The data is provided by Dr. Premysl Landa, Laboratory of Plant Biotechnologies, Institute of Experimental Botany AS CR, Czech Republic.

A significant progress in the field of dispersion electrospinning is the use of water based polymeric formulations made without any use of organic solvents and surfactants. This is called green electrospinning. Special amphiphilic block copolymers with hydrophilic and hydrophobic segments can self-stabilize the dispersion and are used for green electrospinning [46–48]. One of the examples is shown in Scheme 5.9 for the formation of amphiphilic block copolymers of poly(ethylene glycol) with polycaprolactone (MPEG-PCL) and poly(hexamethylene adipate) (MPEG-PHA). PEG is a polar hydrophilic polymer whereas PCL and PHA are hydrophobic polyesters. The block copolymers of two will make a molecular solution in appropriate organic solvent but disperse in the form of nanoparticles in water by making micellar structure.

The electrospinning formulation made up of self-dispersing block copolymer of poly(ethylene glycol) and aliphatic polyester (MPEG-PHA) in water without any surfactant provided no obvious toxic effect on plant growth. This shows a clear advantage of green electrospinning formulations (Figure 5.11).

Dispersions alone are not electrospinnable due to the low viscosity. Here you can imagine the viscosity of milk. Therefore, for electrospinning a template polymer is added in small amounts to the dispersion. The template polymer can be any water soluble polymer. Once the dispersion, together with the template polymer, is spun to fibers, the template polymers can be easily removed if required by putting fiber mats in water without disturbing the morphology of fibers [49].

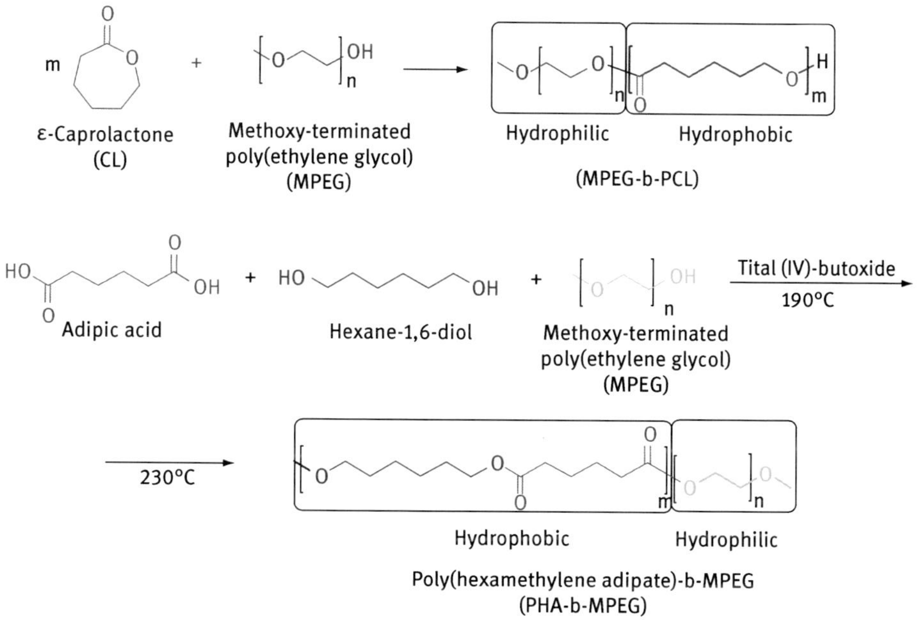

Scheme 5.9: Amphiphilic block copolymers of poly(ethylene glycol)-polycaprolactone (MPEG-PCL) and poly(ethylene glycol)-poly(hexamethylene adipate) (MPEG-PHA) used for green electrospinning.

Polymeric dispersions can be neutral, cationic or anionic depending on the polymeric chemical entities. Therefore, electrospinning of the corresponding charged dispersions can provide fiber mats with positively and negatively charged surfaces. Such mats are highly interesting for properties modification by a layer-by-layer process. Without much effort the fiber mats can be decorated with dyes, nanoparticles or other polymers simply by using another component with the opposite charge (Figure 5.12). This might not be only relevant for agricultural applications but otherwise also for many other areas.

For agricultural applications of electrospun fibers, it would be important to use biodegradable polymers, if possible. The best scenario for use would be when the leftover biodegradable fiber mats after releasing the active components could be degraded in the field without the need of collecting them back. Thereby, soil pollution could be prevented. In one of the studies the green electrospinning formulation based on amphiphilic biodegradable polymer (MPEG-PHA: Scheme 5.9) was used for making carrier fiber mat for pheromones ((E,Z)-7,9-dodecadien-l-yl acetate) encapsulated in biodegradable oligo polylactide microparticles. The resulting mats would act as dispensers for pheromone, disrupting mating of the European grapevine moth *Lobesia botrana* and providing biotechnical plant protection [50].

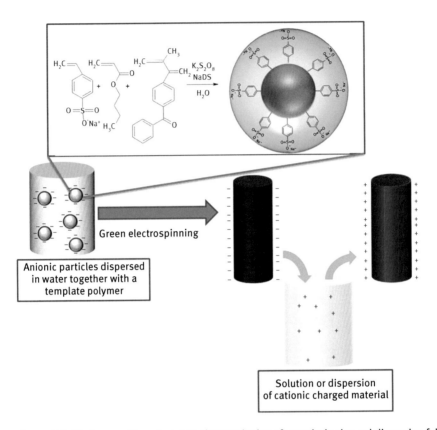

Figure 5.12: Electrospun fibers by green electrospinning of negatively charged dispersion followed by surface modification by layer-by-layer process by simply dipping in a solution/dispersion of opposite charge. [Adapted from *Adv. Funct. Mater.* **2013**, *23*, 3156. Copyright WILEY-VCH Verlag GmbH & Co. KGaA, Weinheim (2013).]

To avoid any confusion, it is worth to clarify that emulsions are different from suspensions/dispersions. An emulsion is a physical mixture of two or more immiscible liquids, such as oil and water. One liquid is present as the dispersed phase and another is present as the continuous phase. The most common example from everyday life is milk and mayonnaise. The simple emulsions can be oil-in-water or water-in-oil type. An oil-in-water emulsion has oil droplets dispersed in water, whereas a water-in-oil emulsion will have water droplets dispersed in a continuous oil phase. Water-in-oil emulsions have been used for electrospinning for incorporating hydrophilic and organic solvent sensitive drugs and biomolecules such as proteins and DNA in hydrophobic fibers. They play an important role in the drug-release applications of electrospun fibers [51–54].

5.2 Hydrogel fibers

A special class of cross-linked polymers are hydrogels [55–57]. They are physically or chemically cross-linked polar polymers which have a very high affinity for water. The non-cross-linked polymer precursors are generally water soluble and can be anionic, cationic or neutral polymers. The cross-linked poly(2-hydroxyethyl methacrylate) and copolymers are used as soft contact lens materials and are one of the oldest hydrogels known. Cross-linked poly(acrylic acid sodium salt) is another common example of a hydrogel being used in large amounts in pampers and hygiene products, also called as superabsorbers. The hydrogels swell with different degrees in water, depending on the chemical structure of the polymer, the cross-linking density and the type of cross-linker used. Moreover, depending on the monomer repeat units / functional groups on the precursor polymer, the hydrogels can show different swelling behaviors at different conditions of pH, temperature, and light. Such hydrogels are called responsive hydrogels. The cross-linked poly(N-isopropylacrylamide) [poly(NIPAm)] is one of the examples of thermoresponsive hydrogels which show decreased swelling on increasing temperature. Such hydrogels are called thermophobic hydrogels [58, 59]. Non cross-linked poly(NIPAm) is a classical example of a polymer showing lower critical solution temperature (LCST), i.e., a temperature-dependent solubility in water with a critical temperature, after which the polymer is insoluble in water as described in Chapter 3. The interpenetrating polymer network of polyacrylic acid and polyacrylamide on the other hand shows an increased swelling with an increase in temperature. They are called thermophilic hydrogels [60, 61]. Other examples of thermophilic hydrogels are cross-linked poly(acrylamide-co-acrylonitrile) and poly(N-acryloylglycinamide) [62, 63]. The corresponding non cross-linked polymers are an example of polymers which show an upper critical solution temperature (UCST) in water. Similarly, pH-responsive hydrogels are also very well-known and the most common examples are cross-linked polyacrylic acid and copolymers, poly(2-hydroxy ethyl methacrylate), poly(dialkyl aminoethyl methacrylate) and poly(vinyl amine). A pH dependent ionization of the pendent groups is responsible for a change in swelling with pH. The pK_a of acid groups decide in which pH range the hydrogel will swell and deswell. For example, the acid groups will be ionized above pK_a and lead to ionic repulsions among macromolecular chains and therefore stronger swelling.

Hydrogels based on biopolymers such as chondroitin sulfate, hyaluronic acid, chitosan, and gelatin are also known. The chemistry of cross-linking and the use of biopolymers for hydrogel applications can be referred from review articles [64].

Hydrogels are interesting materials for biomedical applications such as drug-release encapsulations of biomolecules and wound healing. The hydrogel precursor polymers can be electrospun from water or organic solvents and later cross-linked by heat, light or chemicals. For example, poly(N-vinylpyrrolidone) is a water soluble polymer which can be easily spun from water and subsequently cross-linked to get hydrogel fibers by heating at 200 °C [65]. The use of high temperatures for cross-linking could be problematic, in particular when biomolecules or drugs are encapsulated, as it might lead to their degradation or inactivation. Therefore, milder methods of cross-linking are more suitable such as the utilization of UV – visible light. Polar water soluble polymers, modified with photo cross-linkable groups, together with appropriate photoinitiators can be spun followed by cross-linking by an exposure to light of particular wavelengths to give cross-linked water swellable gel fibers. In one of the examples, the hydrogel precursor polymer (poly(vinyl alcohol), PVOH) was modified to introduce photo-cross-linkable groups by reaction with 3-(2-thienyl)acryloyl chloride which were used for cross-linkings by UV light after fiber formation (Scheme 5.10) [66].

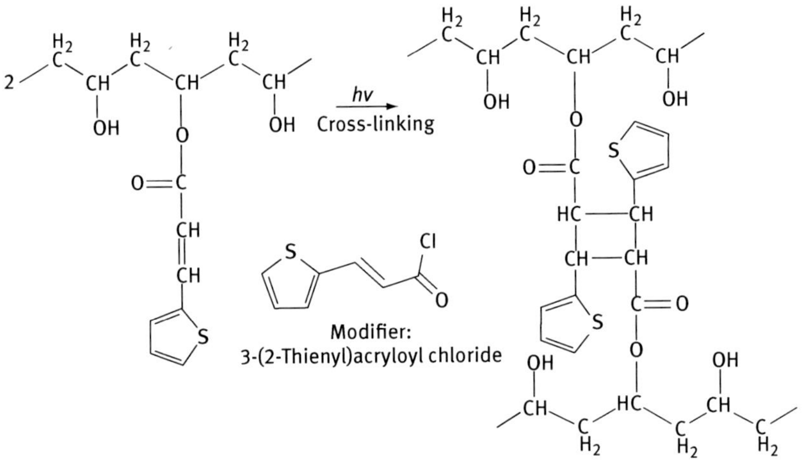

Scheme 5.10: Modified photo cross-linkable poly(vinyl alcohol). Electrospinning followed by UV cross-linking provides water-stable gel fibers.

Also, 4-acryloyloxybenzophenone (ABP) and N-(4-benzoylphenyl) acrylamide (BPAm) are photo cross-linkable monomers. They can be introduced easily to the backbone of hydrogel precursor polymers made by a vinyl polymerization and lead to cross-linking by exposure to UV light as shown recently for making thermoresponsive and thermophilic hydrogel fibers based on copolymers of acrylamide and acrylonitrile (Figure 5.13) [67].

The high porosity of electrospun hydrogel fibers provides a fast diffusion of solvents and therefore enhanced swelling rates in comparison to the bulk hydrogels. The fact was utilized in making the fastest thermoresponsive polymeric actuator, which is a bilayer system made up of thermoplastic polyurethane and photo cross-linked poly (NIPAm) (LCST ~ 29 °C) fibers as two layers. The differential swelling/shrinkage of the two layers with temperature in water provides ultrafast (< 1 s) bending and rolling movements (Figure 5.14).

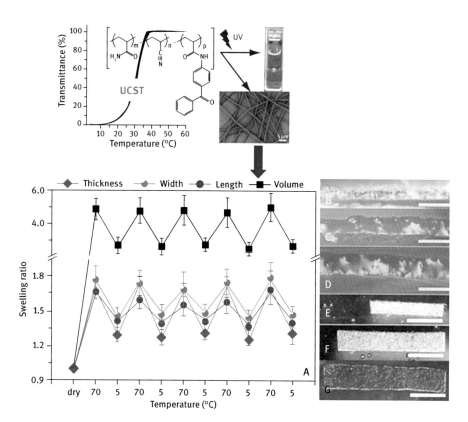

Figure 5.13: A copolymer of acrylamide and acrylonitrile with a photo cross-linker shows UCST behavior in solution. It is electrospinnable and provides hydrogels, in both solution and solid state, on cross-linking (upper picture). The bulk hydrogels and fibers show temperature-dependent swellability in water (lower picture). (A) Temperature-dependent change in thickness, width, length and volume of electrospun fibrous mat; (B, C and D) Microscope photos show the thickness change of a fibrous mat in dry state, at 5 °C and 70 °C, respectively; (E, F, G) Microscope photos of a small piece of nanofiber mat in dry state, wet state at 5 °C and 70 °C. Scale bar of (B, C and D) = 100 μm and scale bar of (E, F and G) = 500 μm. (Reprinted from *Polym. Chem.* **2015**, *6*, 2769–2776, published by Royal Society of Chemistry.)

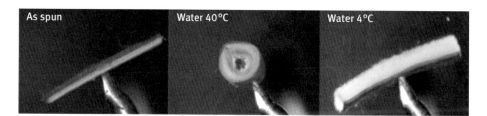

Figure 5.14: Actuation of an electrospun bilayer of photo cross-linked TPU and poly(NIPAm) in water at different temperatures (pink side is TPU). [Reprinted from an open access article *Adv. Mater.* **2015**, DOI:10.1002/adma.201502133. Copyright the Authors. Published by WILEY-VCH Verlag GmbH & Co. KGaA, Weinheim (2015).]

Other chemical reactions are also very often used for cross-linking fibers, which are made from polar polymers providing fibrous hydrogels. A cross-linking agent can be simply mixed with the polymer inside the electrospinning solution. Cross-linked type I collagen fibers were obtained by spinning a solution of collagen, containing the chemical cross-linkers 1-ethyl-3-(3-dimethyl-aminopropyl)-1-carbodiimide hydrochloride (EDC) and N-hydroxysuccinimide (NHS) [68]. In other studies, cross-linked chitosan [69], gelatin [70], and poly(vinyl alcohol) [71] fibers were made by adding genipin and glutaraldehyde, respectively as cross-linking agents. PVOH fibers were also cross-linked by the addition of maleic anhydride [72] and polyacrylic acid [73] followed by heating at high temperatures. The chemistry of cross-linking PVOH is shown in Scheme 5.11.

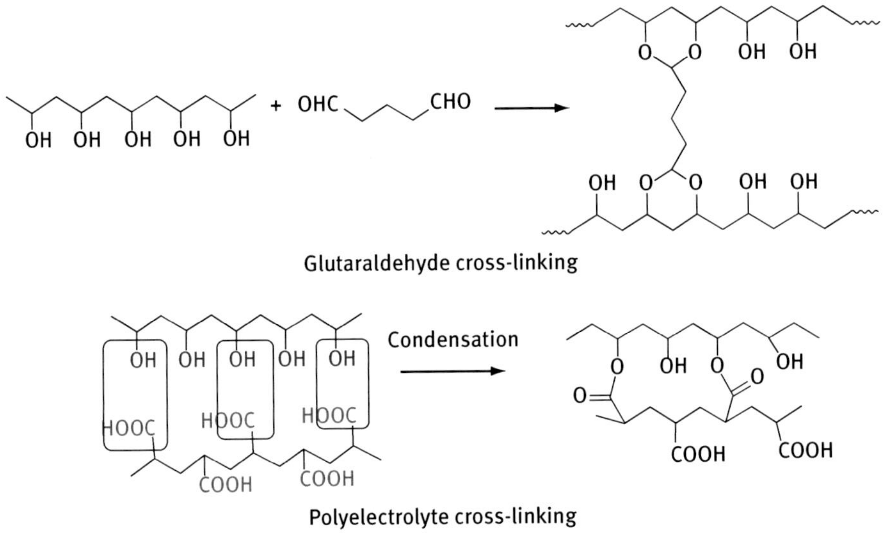

Glutaraldehyde cross-linking

Polyelectrolyte cross-linking

Scheme 5.11: Chemical reactions showing cross-linking of poly(vinyl alcohol) with glutaraldehyde and poly(acrylic acid).

Although this method for making cross-linked fibers is simple, it limits its use as the solution should be freshly made and spun immediately to avoid an unwanted cross-linking in the solution. If the cross-linking starts already in the solution, it will lead to a nonspinnable, gelled material with time. In addition, there is always a narrow window for the ratio of cross-linker and matrix polymer for getting spinnable solutions with appreciable stabilities for spinning. This requires lot of optimizations in order to avoid gel formations before spinning.

The same chemistry of chemical cross-linking can be used in a two-step process, i.e., first make nanofibers and then cross-link them in the second step. For example,

PVOH fibers were cross-linked by dipping them in glutaraldehyde or poly(ethylene glycol) (PEG) diacylchloride solutions [74]. Chloromethylated polysulfone membranes were also cross-linked by a soaking in diamine or diol solutions [75, 76].

For a particular polymer, there could be different cross-linking options. It is very important to choose the right cross-linking method in terms of cross-linking agent and cross-linking conditions depending on the intended application. For example, hydrogels intended as scaffolds for tissue engineering should be cross-linked under mild conditions (not too high temperatures and no use of toxic UV light that can kill cells), using nontoxic chemicals and solvents. Let us take the example of gelatine, a denatured collagen which is a very well-known biodegradable and biobased biopolymer, suitable for the use as scaffolds for tissue engineering besides many other applications. It is soluble in fluorinated solvents such as 1,1,1,3,3,3-hexaflouro-2-propanol (HFIP) or 2,2,2-triflourothanol (TFE), which could be toxic to cells. This makes them unsuitable for electrospinning if the fibers are intended for a scaffold application. Less toxic solvent mixtures, such as acetic acid/ethyl alcohol or acetate/water, could be an option for electrospinning of gelatin for this purpose [77]. The choice of an appropriate cross-linker is another important consideration for making water-insoluble scaffolds. Gelatine can be cross-linked with glutaraldehyde, glyceraldehyde, genipin, reactive oxygen species created by plasma treatment, and glucose. In this case glyceraldehyde, genipin, and glucose represent less toxic alternatives for cross-linking [78]. Alginate fibers can be cross-linked by an ionic cross-linking in a post-treatment, by dipping them in a CaCl$_2$ solution [79].

The electrospinning of already cross-linked hydrogels is possible in the form of microparticles. They can be dispersed in a template polymer solution and electrospun as shown for microparticles of a copolymer of N-isopropylacrylamide and acrylic acid (AAc), cross-linked with N,N-methylenebis (acrylamide). The particles were dispersed in a DMF solution of poly(vinyl pyrrolidone) and electrospun. The presence of acrylic acid and N-isopropylacrylamide provided pH and temperature sensitivity to the microparticles as shown in Figure 5.15 [65].

The method is not limited to these specific microparticles but can also be applied to other cross-linked hydrogel particles. In fact, highly concentrated microparticle solutions of N,N-dimethylacetylamide and 4-vinylpyridine (VP)/acrylamidoethylsulfonic acid, cross-linked with ethylene dimethacrylate (EDMA) in water or ethanol, provided fibers by electrospinning without the use of any template polymer [80].

There are different methods for making microgel particles, starting from the corresponding monomer and a vinyl cross-linker (note: monomers with a minimum of two polymerizable double bonds lead to cross-linking during polymerization and are called cross-linkers) such as emulsion polymerization (oil-in-water, water-in-oil), miniemulsion, suspension and precipitation polymerization.

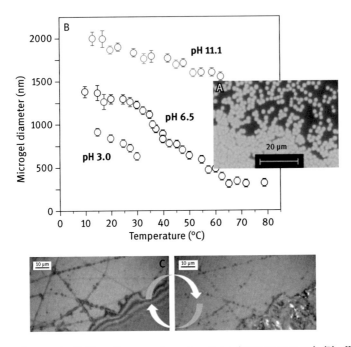

Figure 5.15: (A) Monodisperse microgel particles (diameter 1.8 μm); (B) effect of temperature on the microgel diameter at different pH; (C) reversible change in the particle diameter with temperature at pH 11; at 25 °C particles are swollen and at 45 °C they are deswollen. [Reprinted with permission from *Macromol. Rapid Commun.* **2010**, *31*, 183–189. Copyright Wiley-VCH Verlag GmbH & Co. KGaH Weinheim (2010).]

5.3 Electrospinning of polymers with complex architectures

The macromolecular chains in a polymer are not limited to linear structures but can also be branched, stars, combs, brushes, rings or dendrimers, etc. (Figure 5.16).

The polymer architecture affects the solution viscosity, melt viscosity and solubility. These are important parameters for electrospinning. For similar molar masses, the entanglement concentration (C_e) is higher for branched polymers as branching hinders chain entanglements and overlaps. (Table 5.3) [81].

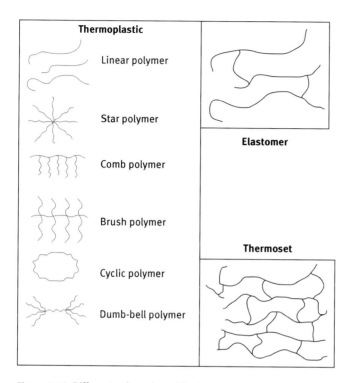

Figure 5.16: Different polymeric architectures.

Table 5.3. Effect of branching on the entanglement concentration of poly(ethylene terephthalate-co-ethylene isophthalate).

mol% Branching point	M_w (kg/mol)	M_w/M_n	Ce (kg/mol)
0	54.2	1.7	6.0
0.5	49.2	2.4	7.0
0	77.3	1.6	4.5
0.5	94	2.9	5.5
3.0	76	10.7	10

Adapted with permission from *Macromolecules* **2004**, *37*, 1760. Copyright American Chemical Society (2004).

Aside from that, polymers can have either only one type of repeat unit (homopolymers) or can have two or more than two types of repeat unit structures covalently linked to each other (copolymers). Copolymers are different to polymer blends in which two or more than two polymers are physically mixed. The arrangement of co-monomeric units can be statistical, alternate or in the form of blocks in a copolymer. Therefore, the corresponding polymers are called statistical, alternate or block copolymers (Figure 5.17).

A copolymer which has two monomeric units – A and B – is represented as poly(A-co-B), poly(A-alt-B), and poly(A-b-B), for statistical, alternate and block copolymers, respectively. The polymer properties can be modified easily by making copolymers.

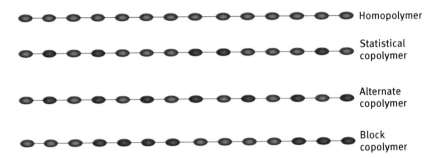

Figure 5.17: Monomer repeat unit arrangement in homo- and various copolymers.

The choice of a homopolymer for electrospinning is commanded by the application. Conducting polymers, such as polyaniline, poly(3,4-ethylenedioxythiophene), polypyrrole, and poly(phenylene vinylene) with appropriate dopants, are suitable for making fibrous nonwoven mats for the use as sensors and diodes. Besides homopolymers, large numbers of copolymers have been spun due to different purposes. The choice of co-monomers also depends on the application. A photo cross-linkable statistical copolymer of 3-acrylamidophenylboronic acid (AAPBA) and 2-hydroxyethylmethacrylate (HEMA) was electrospun to get pH and glucose responsive nanofibers for a reversible binding of lectins [82].

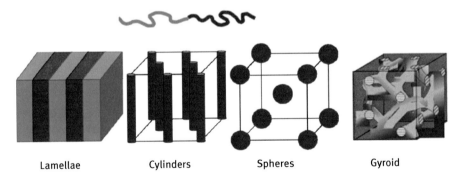

Figure 5.18: Morphological change on varying the volume fraction (f) of one of the blocks in a block copolymer. [Adapted from *Int. J. Mol. Sci.* **2009**, *10*(9), 3671–3712, an open access article published by MDPI AG, Basel, Switzerland.]

Block copolymers are a highly interesting type of materials as they phase separate into different morphologies, depending on the block lengths and interaction between the two blocks [83]. Figure 5.18 shows some important block copolymer morphologies, depending on the fraction of block A for AB block copolymers.

The state-of-the art method for making block copolymers is by anionic polymerization, which is also called living polymerization. The name *living* signifies active chain-ends on each macromolecule during polymerization due to the absence of transfer and termination reactions. Radical anions, carbanions and oxyanions are some of the initiators for anionic polymerizations. In an ideal case, the rate of chain initiation is much faster than the rate of propagation. This leads to the start of all chains at the same time without any termination or transfer reaction occurring during the polymerization. This results into a polymer with a desired molar mass (depends on monomer to initiator ratio), a very low-molar-mass distribution (around 1.0–1.1) and defined chain-ends. This method is the state-of-the-art method for making block copolymers by a sequential addition of two monomers, i.e., once the polymerization of the first monomer is finished, the second monomer is added (Scheme 5.12).

By this method many different vinyl monomers can be polymerized sequentially. It is a technical method of making many commercially available block copolymers such as poly(styrene-b-butadinene-b-styrene) or poly(styrene-b-isoprene-b-styrene) block copolymers. Block copolymers are of interest for unique combinations of properties (for example water soluble blocks with water-insouble block, which results in amphiphilic polymers with tenside properties in solutions), self-assembly, and the use as compatibilizers for incompatible polymers in a blend.

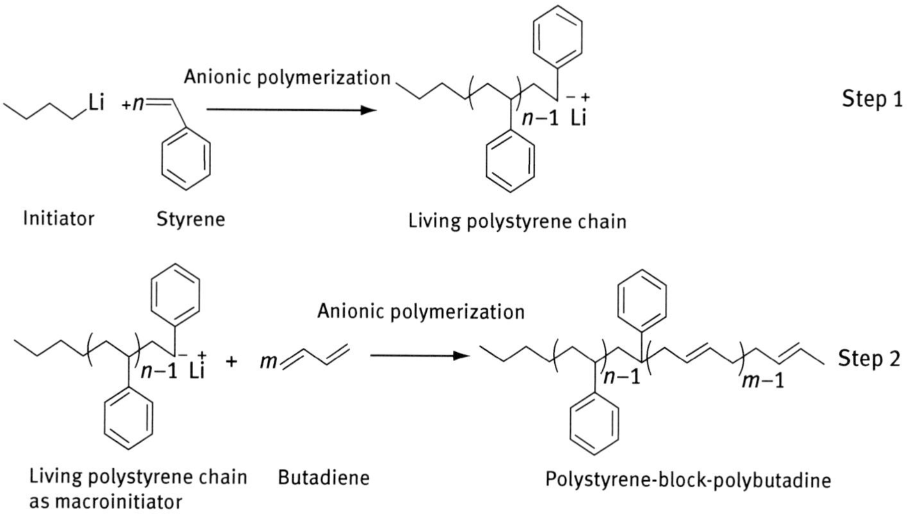

Scheme 5.12: Synthetic scheme for the formation of AB block copolymer of styrene and butadiene by a sequential anionic polymerization. For simplicity, the 1,4 polymerization of butadiene is shown. It can also undergo a 1,2 addition, leading to double bonds in the side chain. Furthermore, 1,4 polybutadiene can have a *cis* or *trans* configuration.

Electrospun fibers are interesting 1D nanostructures for studying the self-assembly behavior of block copolymers in nano-confinements and for generating novel morphologies. Of special interest is coaxial electrospinning in which block copolymers can be encapsulated in a thermally stable shell for developing different self-assembled structures by annealing (Figure 5.19) [84].

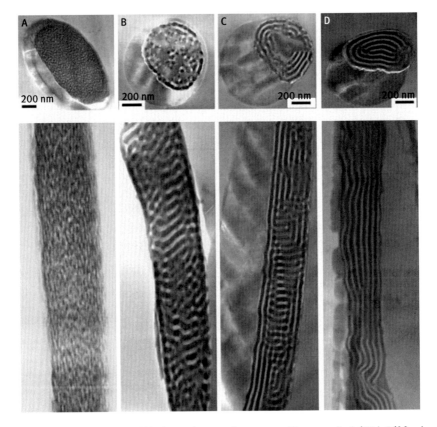

Figure 5.19: TEM images of block copolymers of styrene and isoprene [poly(PS-b-PI)] [molar masses of PS and PI blocks were 74,500 and 68,400 g/mol, respectively (53 vol% of polyisoprene)]. The block copolymers were electrospun as core with a SiO$_2$ shell. (A) As-spun morphology; (B – D) thermally annealed morphologies: (B) stacked lamellar structures, annealing at 125 °C for 24 h; (C) a concentric-cylinder morphology, annealing at 175 °C for 24 h; (D) a parallel morphology by annealing at 175 °C for 50 h. The cross sections – normal and parallel to the fiber axes – are shown in top and bottom rows, respectively. Stained PI and PS domains appear as dark and light portions, respectively. [Reprinted by permission from. Copyright Wiley-VCH Verlag GmbH & Co. KGaH Weinheim (2006).]

The block copolymer architecture is also useful for making functional fibers with properties such as thermo- and pH responsivities in water, for example. Normally, thermo- and pH-responsive polymers show external stimuli (temperature- and pH-) dependent water solubility. Therefore, one of the ways for getting responsive fibers which maintain the fiber form under all conditions is to spin the corresponding polymers and then cross-link them using light or heat. To avoid an extra cross-linking step, block copolymers made up of a responsive polymer block and a hydrophobic block can be used for the same purpose. The hydrophobic block is responsible for maintaining the fiber form and the responsive block changes its conformation with external stimuli. For example, amphiphilic ABA triblock copolymers in which the middle block (B) was either pH responsive [poly(2-diethylamino ethyl methacrylate)] or temperature responsive [poly(N-isopropylacrylamide)] and hydrophobic side blocks (A) such as poly(methyl methacrylate) and polystyrene were electrospun [85, 86].

Graft copolymers can also be used for the same purpose. A graft copolymer architecture is a special case of branched polymers in which the main chain polymer has side-chain branches from some other structurally different polymer. In one of the studies, chitosan-grafted poly(N-isopropylacrylamide) was electrospun to get fibers for a pH and temperature controlled release of drugs [87].

An appropriate choice of the polymers in graft copolymers can also provide special morphologies in electrospun fibers in a simple way. Polyacrylonitrile-graft-poly(dimethyl siloxane) represents an amphiphilic graft copolymer, which undergoes self-assembly in solution. The electrospinning of self-assembled solution provides highly porous fibers [88]. Other methods of making porous fibers using either highly volatile solvents or porogens such as salts and soluble/pyrolysable polymeric additive are described in Chapter 2.

5.4 Chemistry on electrospun fibers

Until now we talked about the electrospinning of different types of polymers and polymeric architectures. Electrospun fibers can be modified on the surface by *grafting-from*, *grafting-to*, and *grafting through* techniques as well, for introducing additional functionalities, using an appropriate chemistry. The surface modification of nanofibers brings extra advantages in terms of changing the hydrophilicity – hydrophobicity, affinity for particular molecules (affinity membranes) or responsive characters, etc. Before we understand the chemistry of nanofibers, let us revise in short the techniques of polymerizations. The process of formation of polymers is classified as chain or step polymerization. Chain polymerization is generally carried out for vinyl monomers with double bonds. The initiators for chain polymerization are mainly classified into four categories: radical, anionic, cationic and metal catalysts. The corresponding polymerization techniques are called radical, anionic, cationic and metal-catalyzed polymerizations. Azo compounds and organic/inorganic peroxides generate radicals at different temperatures, depending on the structure. They are the most commonly used thermal initiators in radical polymerizations. Very high-molar-mass polymers can be easily made by radical polymerizations with a molar mass dispersity (Đ) of 1.5–2.0. Radicals can also be generated by a cleavage of bonds by UV, visible light or gamma irradiation and under mild conditions by redox reactions. The Ce(IV)-alcohol redox polymerization or other redox initiators can be simply applied for a modification of nanofiber surface with OH groups. Either OH-group carrying polymers can be first spun and then modified via a Ce(IV)-induced graft polymerization of vinyl monomers or OH groups may be generated on fiber surface using a plasma discharge followed by surface grafting via redox polymerization.

Modified radical polymerization methods were introduced in the last years such as atom transfer radical polymerization (ATRP), reversible addition-fragmentation chain – transfer polymerization (RAFT) and nitroxide mediate polymerization (NMP). They are known with different names in the literature such as, controlled radical polymerization, pseudo/quasi living polymerization, and reversible decativation radical polymerization. They offer the advantage of minimized terminations and transfer reactions, which are otherwise typical for conventional radical polymerizations. Thereby, they provide predicted molar masses, low-molar-mass dispersities and control over chain-ends and have been used in the last two decades not only for making different macromolecular architectures like block or star polymers but also for modifying surfaces by a grafting with appropriate polymers. Such methods could also be highly interesting for modifying electrospun fibers for specific purposes (Figure 5.20). For example, a very fast surface modification of electrospun fibers with the thermoresponsive polymer poly(N-isopropylacrylamide) was shown, using a controlled radical polymerization from the fiber surface. An ATRP initiator was incorporated in the polymer for spinning in the form of a co-monomer and subsequently the ATRP-initiator sites were used for the polymerization of the vinyl monomer (N-isopropylacrylamide) on the fiber surface. So the process is a *grafting-from* approach. Similar methods can be applied for immobilizing RAFT chain-transfer agents onto electrospun fibers for further modifications [89, 90].

A *grafting-to* approach uses readymade polymeric chains with suitable functional groups for reactions with active/appropriate sites on fiber surfaces. This means we click an already made polymer chain on the fiber surface. There are different ways for clicking a functional moiety/polymeric chain to a fiber surface. For example, fibers which have activated ester units (pentafluorophenyl, o-nitrobenzyl) can be utilized for reactions with nucleophilic chain-ends such as amine groups. Copper catalyzed alkyne-azide 1,3-cycloaddition is another method for clicking two molecules, containing alkyne and azide groups via the formation of a 1,2,3-triazole connecting ring. Electrospun fibers can have either alkyne groups or azide groups which can be clicked with other molecules on the surface with appropriate chain-ends. For example, thermoresponsive and photoresponsive fiber surfaces could be generated by clicking poly(NIPAm) and azobenzene units with alkyne chain-ends onto polymeric fibers with azide groups. The same reaction can also be used for clicking bioactive peptides, fluorescent dyes, functional polymers, etc., in order to enhance the utility areas of nanofibers [91,92]. Thiol-ene chemistry is also getting more and more attention for modifying fiber surfaces [93].

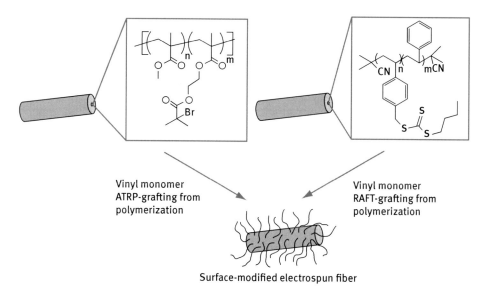

Figure 5.20: *Grafting-from* method for modifying the surfaces of electrospun fibers. Vinyl polymers with either ATRP initiating sites or RAFT transfer agents are electrospun and later modified by polymerizing appropriate monomers on the fiber surface.

In one of the studies, a cell-adhesive peptide was conjugated to the hydrophilic block poly[oligo(ethylene glycol) methacrylate] (POEGMA) in a block copolymer fiber of POEGMA-polylacide, using thiol-ene chemistry [94]. The details about different approaches used for click reactions on electrospun fibers can be seen in a recent review article [95].

Moreover, the electrospun fibers can be spun with initiators and catalysts for further modification by simple mixing in the spinning solution. For example, FeCl$_3$ was mixed in polystyrene spinning solution for subsequent surface polymerization of pyrrole on the surface of polystyrene fibers for introducing conductivity [96].

Conductive metal films can also be coated on electrospun fibers by spinning a polymer solution with metal salt, for example, HAuCl$_4$·3H$_2$O, for gold deposition. This is followed by a reduction of the metal salt to give gold particles. The last step would be to use immobilized gold particles for catalyzing the reduction of gold salt by hydroxylamine to get a continuous metal film [97].

5.5 Interesting to know

– Poly(vinyl alcohol) (PVOH) and *cis*-polyisoprene are rare examples of biodegradable vinyl polymers with a C–C backbone. Although C–C linkages can be thermally degraded at high temperatures (pyrolysis), they are highly stable against hydrolysis and enzymatic degradations. Therefore, vinyl polymers with a C–C backbone such as polyethylene, polypropylene, poly(methyl methacrylate), polystyrene, poly(vinyl chloride) and poly(vinyl acetate) are not categorized as biodegradable polymers. PVOH is degraded by suitably acclimated microorganisms, such as several species of Pseudomonas. The tertiary-carbon in PVOH undergoes an oxidation by specific oxidases and dehydrogenases, providing hydrolysable β-hydroxylketones and 1,3-diketones on the polymer backbone (Scheme 5.13). Hydrolysis of 1,3-diketones in a polymer backbone by β-ketonehydrolase provides oligomers with methyl ketone and acid chain-ends, which are metabolized by the Krebs cycle in microorganisms [98].

– Cis-1,4-polyisoprene (PI) is the major constituent of natural rubber, obtained from the *Hevea brasiliensis* (rubber tree). Natural and synthetic PI are degraded by several species of *Actinomycetes* (e.g., *Streptomyces*, *Nocardia*, *Micromonospora* and *Actinoplanes*), *Pseudomonas aeruginosa*, *Pseudomonas citronellolis* by an oxidative cleavage at the double bond, providing low-molecular-weight oligo(*cis*-1,4-isoprene) molecules with aldehyde and keto groups at their ends for a metabolism by microorganisms [99, 100].

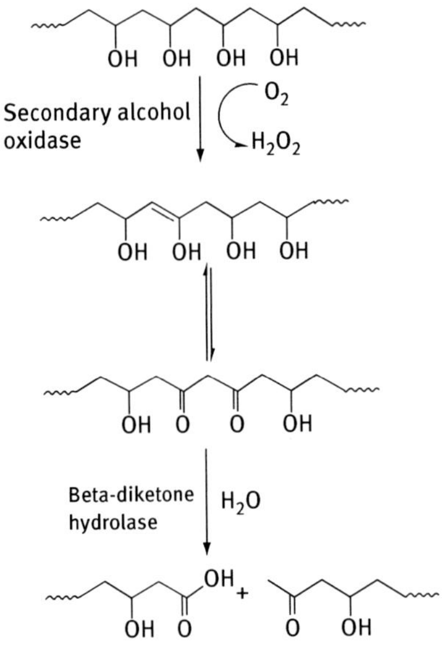

Scheme 5.13: Enzymatic cleavage of C–C backbones in poly(vinyl alcohol). Refer to Ref. 95 for details about degradation mechanism.

- Aromatic polyesters, for example, poly(ethylene terephthalate) (PET), poly(buty-lene terephthalate), with aromatic moieties in each repeat unit, are not biode-gradable. The rigidity of the polymer chain in addition to its chemical structure plays an important role in its susceptibility to enzymatic attacks. H-bonded polyamides (nylons) are also not biodegradable due to very high rigidities. On the other hand, aliphatic-aromatic polyesters can be biodegraded, depending on the amount of aromatic moieties, hydrophilicity and crystallinity. The statis-tical copolymers [poly(trimethylene decanedioate-co-trimethylene terephtha-late)] made by a condensation of 1,3-propanediol, decanedioic acid and ter-ephthalic acid were biodegradable if the amount of aromatic units (terephthalic acid) was kept low (around 10 mol%) [101].

Furthermore, PET modified with PEG made by a condensation of hydroxyl-func-tionalized PEG, ethylene glycol and terephthalic acid is a hydrolytically and enzymatically degradable aliphatic – aromatic polyester due to an increased hydrophilicity [102]. Ecoflex is one of the commercially available aliphatic-aro-matic compostable polyesters made by the condensation of butanediol, terephtha-lic acid and adipic acid.

References

[1] Q. Hu, H. Wu, L. Zhang, H. Fong, M. Tian, *Express Polym. Lett.* **2012**, *6*, 258–265.

[2] X. Xu, J.-F. Zhang, Y. Fan, *Biomacromolecules* **2010**, *11*, 2283–2289.

[3] R. Wu, J.-F. Zhang, Y. Fan, D. Stoute, T. Lallier, X. Xu, *Biomed. Mater. (Bristol, England)* **2011**, *6*, 035004.

[4] Y. Ji, K. Ghosh, B. Li, J. C. Sokolov, A. F. R. Clark, M. H. Rafailovich, *Macromol. Biosci.* **2006**, *6*, 811–817.

[5] H. Ma, C. Burger, B. S. Hsiao, B. Chu, *J. Mater. Chem.* **2011**, *21*, 7507.

[6] P. Heikkilä, A. Taipale, M. Lehtimäki, A. Harlin, *Polym. Eng. Sci.* **2008**, *48*, 1168–1176.

[7] R. Gopal, S. Kaur, C. Y. Feng, C. Chan, S. Ramakrishna, S. Tabe, T. Matsuura, *J. Membr. Sci.* **2007**, *289*, 210–219.

[8] D. Aussawasathien, C. Teerawattananon, A. Vongachariya, *J. Membr. Sci.* **2008**, *315*, 11–19.

[9] K. Yoon, K. Kim, X. Wang, D. Fang, B. S. Hsiao, B. Chu, *Polymer* **2006**, *47*, 2434–2441.

[10] A. Chaudhary, A. Gupta, R. B. Mathur, S. R. Dhakate, *Adv. Mat. Lett.* **2014**, *5*, 562–568.

[11] B.Veleirinho, J. A.Lopes-da-Silva, *Process Biochem.* **2009**, *44*, 353–356.

[12] N. Ashammakhi, A. Ndreu, A. Piras, L. Nikkola, T. Sindelar, H. Ylikauppila, A. Harlin, E. Chiellini, V. Hasirci, H. Redl, *J. Nanosci. Nanotech.* **2006**, *6*, 2693–2711.

[13] R. W. Lenz, R. H. Marchessault, *Biomacromolecules* **2005**, *6*, 1–8.

[14] K. Leja, G. Lewandowicz, *Polish J. Environ. Stud.* **2010**, *19*, 255–266.

[15] L. S. Nair, C. T. Laurencin, *Prog. Polym. Sci.* **2007**, *32*, 762–798.

[16] S. A. Riboldi, M. Sampaolesi, P. Neuenschwander, G. Cossu, S. Mantero, *Biomaterials* **2005**, *26*, 4606–4615.

[17] A.Gugerell, J.Kober, T.Laube, T.Walter, S.Nürnberger, E.Grönniger, S.Brönneke, R.Wyrwa, M.Schnabelrauch, M.Keck, *PloS One* **2014**, *9*, e90676.

[18] S. Agarwal, J. H. Wendorff, A. Greiner, *Adv. Mater. (Deerfield Beach, Fla.)* **2009**, *21*, 3343–3351.

[19] D. R. Nisbet, A. E. Rodda, M. K. Horne, J. S. Forsythe, D. I. Finkelstein, *Biomaterials* **2009**, *30*, 4573–4580.

[20] F. Yang, R. Murugan, S. Wang, S. Ramakrishna, *Biomaterials* **2005**, *26*, 2603–2610.

[21] C. Y. Xu, R. Inai, M. Kotaki, S. Ramakrishna, *Biomaterials* **2004**, *25*, 877–886.

[22] C. M. Murphy, F. J. O'Brien, *Cell Adhesion & Migration* **2014**, *4*, 377–381.

[23] A. K. Ekaputra, G. D. Prestwich, S. M. Cool, D. W. Hutmacher, *Biomacromolecules* **2008**, *9*, 2097–2103.

[24] S. M. Peltola, F. P. W. Melchels, D. W. Grijpma, M. Kellomäki, *Ann. Med.* **2008**, *40*, 268–280.

[25] L.Moroni, R.Schotel, D.Hamann, J. R.deWijn, C. A.vanBlitterswijk, *Adv. Funct. Mater.* **2008**, *18*, 53–60.

[26] X. Yang, J. D. Shah, H. Wang, *Tissue Engineering. Part A* **2009**, *15*, 945–956.

[27] J. K. Kang, M. H. Lee, B. J. Kwon, H. H. Kim, I. K. Shim, M. R. Jung, S. J. Lee, J.-C. Park, *Macromol. Res.* **2012**, *20*, 795–799.

[28] G.Duan, S.Jiang, V.Jérôme, J. H.Wendorff, A.Fathi, J.Uhm, V.Altstädt, M.Herling, J.Breu, R.Freitaget al., *Adv. Funct. Mater.* **2015**, *25*, 2850–2856; Y. Si, J. Yu, X. Tang, J. Ge, B. Ding, *Nature Commun.* **2014**, doi: 10.1038/ncomms6802.

[29] X. Hu, S. Liu, G. Zhou, Y. Huang, Z. Xie, X. Jing, *J. Contr. Release: Official J. Contr. Release Soc.* **2014**, *185*, 12–21.

[30] Y. Zhang, S. Sinha-Ray, A. L. Yarin, *J. Mater. Chem.* **2011**, *21*, 8269.

[31] S. Demirci, A. Celebioglu, Z. Aytac, T. Uyar, *Polym. Chem.* **2014**, *5*, 2050–2056.

[32] B. Ding, M. Yamazaki, S. Shiratori, *Sensors and Actuators B: Chemical* **2005**, *106*, 477–483.

[33] B.Ding, M.Kikuchi, C.Li, S.Shiratori in *Nanotechnology at the leading edge. Chapter 1. Electrospun Nanofibrous Polyelectrolytes Membranes as High Sensitive Coatings for QCM-Based Gas Sensors* (Ed.: E. V.Dirote), Nova Science Publishers, New York, **2006**.

[34] A. Kumar, R. Jose, K. Fujihara, J. Wang, S. Ramakrishna, *Chem. Mater.* **2007**, *19*, 6536–6542.

[35] S. Chen, H. Hou, F. Harnisch, S. A. Patil, A. A. Carmona-Martinez, S. Agarwal, Y. Zhang, S. Sinha-Ray, A. L. Yarin, A. Greiner, U. Schröder, *Energy Environ. Sci.* **2011**, *4*, 1417.

[36] S. Chen, D. Han, H. Hou, *Polym. Adv. Technol.* **2011**, *22*, 295–303.

[37] S. Jiang, H. Hou, A. Greiner, S. Agarwal, *ACS Appl. Mater Interfaces* **2012**, *4*, 2597–2603.

[38] S. Jiang, G. Duan, J. Schöbel, S. Agarwal, A. Greiner, *Composites Science and Technology* **2013**, *88*, 57–61.

[39] C. J. Luo, E. Stride, S. Stoyanov, E. Pelan, M. Edirisinghe, *J. Polym. Res.* **2011**, *18*, 2515–2522.

[40] G. C. Pontelli, R. P. Reolon, A. K. Alves, F. A. Berutti, C. P. Bergmann, *Applied Catalysis A: General* **2011**, *405*, 79–83.

[41] W. T. Gibbons, T. H. Liu, K. J. Gaskell, G. S. Jackson, *Applied Catalysis B: Environmental* **2014**, *160 – 161*, 465–479.

[42] S. Wen, M. Liang, R. Zou, Z. Wang, D. Yue, L. Liu, *RSC Adv.* **2015**, *5*, 41513–41519.

[43] C. L.Altan, J. M.Jos Lenders, H. H.Paul, G.de With, H.Friedrich, S.Bucak, A. J. M.Nico Sommerdijk, *Chem. Eur. J.* **2015**, *21*, 6150–6156.

[44] A. Greiner, S. Mang, O. Schäfer, P. Simon, *Acta Polym.* **1997**, *48*, 1–15.

[45] F. Mitschang, H. Schmalz, S. Agarwal, A. Greiner, *Angewandte Chemie (International ed. in English)* **2014**, *53*, 4972–4975.

[46] S. Agarwal, A. Greiner, *Polym. Adv. Technol.* **2011**, *22*, 372–378.

[47] K. Bubel, Y. Zhang, Y. Assem, S. Agarwal, A. Greiner, *Macromolecules* **2013**, *46*, 7034–7042.

[48] J. Sun, K. Bubel, F. Chen, T. Kissel, S. Agarwal, A. Greiner, *Macromol. Rapid Commun.* **2010**, *31*, 2077–2083.

[49] A. Stoiljkovic, R. Venkatesh, E. Klimov, V. Raman, J. H. Wendorff, A. Greiner, *Macromolecules* **2009**, *42*, 6147–6151.

[50] P. Bansal, K. Bubel, S. Agarwal, A. Greiner, *Biomacromolecules* **2012**, *13*, 439–444.

[51] S. Agarwal, A. Greiner, *Polym. Adv. Technol.* **2011**, *22*, 372–378.

[52] C. Wang, S. N. Tong, Y. H. Tse, M. Wang, *AMR* **2011**, *410*, 118–121.

[53] T.Briggs, T. L.Arinzeh, *J. Biomed. Mater. Res A* **2014**, *102*, 674–684.

[54] A. Camerlo, C. Vebert-Nardin, R. M. Rossi, A.-M. Popa, *Eur. Polym. J.* **2013**, *49*, 3806–3813.

[55] E. M. Ahmed, *J. Adv. Res.* **2015**, *6*, 105–121.

[56] W. A.Laftah, S.Hashim, A. N.Ibrahim, *Polymer-Plastics Technology and Engineering* **2011**, *50*, 1475–1486.

[57] I. Gibas, H. Janik, *Chemistry & Chemical Technology* **2010**, *4*, 297–304.

[58] N. Nath, A. Chilkoti, *Adv. Mater.* **2002**, *14*, 1243–1247.

[59] E. Gil, S. Hudson, *Prog. Polym. Sci.* **2004**, *29*, 1173–1222.

[60] E. A.Bekturov, L. A.Bimendina in *Advances in Polymer Science* (Eds.: H.-J.Cantow, G.Dall'Asta, K.Dušek, J. D.Ferry, H.Fujita, M.Gordon, J. P.Kennedy, W.Kern, S.Okamura, C. G.Overbergeret al.), Springer Berlin Heidelberg, Berlin, Heidelberg, **1981**.

[61] H. Katono, A. Maruyama, K. Sanui, N. Ogata, T. Okano, Y. Sakurai, *J. Controlled Release* **1991**, *16*, 215–227.

[62] J. Seuring, S. Agarwal, *Macromol. Rapid Commun.* **2012**, *33*, 1898–1920.

[63] J. Seuring, S. Agarwal, *ACS Macro Lett.* **2013**, *2*, 597–600.

[64] S. van Vlierberghe, P. Dubruel, E. Schacht, *Biomacromolecules* **2011**, *12*, 1387–1408.

[65] J. E. Díaz, A. Barrero, M. Márquez, A. Fernández-Nieves, I. G. Loscertales, *Macromol. Rapid Commun.* **2010**, *31*, 183–189.

[66] J. Zeng, H. Hou, J. H. Wendorff, A. Greiner, *Macromol. Rapid Commun.* **2005**, *26*, 1557–1562.

[67] F. Liu, S. Jiang, L. Ionov, S. Agarwal, *Polym. Chem.* **2015**, *6*, 2769–2776.

[68] L. Meng, O. Arnoult, M. Smith, G. E. Wnek, *J. Mater. Chem.* **2012**, *22*, 19412.

[69] E.Mirzaei, R.Faridi-Majidi, M. A.Shokrgozar, F. A.Paskiabi, *Nanomed. J.* **2014**, *1*, 137–146.

[70] T.-H. Nguyen, B.-T. Lee, *JBiSE* **2010**, *03*, 1117–1124.

[71] C.Tang, C. D.Saquing, J. R.Harding, S. A.Khan, *Macromolecules* **2010**, *43*, 630–637.

[72] E. Yang, X. Qin, S. Wang, *Mater. Lett.* **2008**, *62*, 3555–3557.

[73] L. Li, Y.-L. Hsieh, *Nanotechnology* **2005**, *16*, 2852–2860.

[74] Y.Wang, Y.-L.Hsieh, *J. Appl. Polym. Sci.* **2010**, NA.

[75] A. M. Park, F. E. Turley, R. J. Wycisk, P. N. Pintauro, *Macromolecules* **2014**, *47*, 227–235.

[76] A.Park, F.Turley, R.Wycisk, P.Pintauro, *J. Electrochem. Soc.* **2015**, *162*, F560.

[77] K. Sisson, C. Zhang, M. C. Farach-Carson, D. B. Chase, J. F. Rabolt, *Biomacromolecules* **2009**, *10*, 1675–1680.

[78] K. Siimon, H. Siimon, M. Järvekülg, *J. Mater. Sci. Mater Med.* **2015**, *26*, 5375.

[79] C. A. Bonino, M. D. Krebs, C. D. Saquing, S. I. Jeong, K. L. Shearer, E. Alsberg, S. A. Khan, *Carbohydrate Polym.* **2011**, *85*, 111–119.

[80] S. Piperno, L. A. Gheber, P. Canton, A. Pich, G. Dvorakova, A. Biffis, *Polymer* **2009**, *50*, 6193–6197.

[81] M. G. McKee, G. L. Wilkes, R. H. Colby, T. E. Long, *Macromolecules* **2004**, *37*, 1760–1767.

[82] Y. Wang, Y. Kotsuchibashi, K. Uto, M. Ebara, T. Aoyagi, Y. Liu, R. Narain, *Biomater. Sci.* **2015**, *3*, 152–162.

[83] Y. Mai, A. Eisenberg, *Chem. Soc. Rev.* **2012**, *41*, 5969–5985.

[84] V. Kalra, S. Mendez, J. H. Lee, H. Nguyen, M. Marquez, Y. L. Joo, *Adv. Mater.* **2006**, *18*, 3299–3303.

[85] L. Wang, P. D. Topham, O. O. Mykhaylyk, J. R. Howse, W. Bras, R. A. L. Jones, A. J. Ryan, *Adv. Mater.* **2007**, *19*, 3544–3548.

[86] A. Nykänen, S.-P. Hirvonen, H. Tenhu, R. Mezzenga, J. Ruokolainen, *Polym. Int.* **2014**, *63*, 37–43.

[87] H. Yuan, B. Li, K. Liang, X. Lou, Y. Zhang, *Biomed. Mater. (Bristol, England)* **2014**, *9*, 055001.

[88] G. M. Bayley, P. E. Mallon, *Polymer* **2012**, *53*, 5523–5539.

[89] C. Brandl, A. Greiner, S. Agarwal, *Macromol. Mater. Eng.* **2011**, *296*, 858–864.

[90] T. Ameringer, F. Ercole, K. M. Tsang, B. R. Coad, X. Hou, A. Rodda, D. R. Nisbet, H. Thissen, R. A. Evans, L. Meagher, et al., *Biointerphases* **2013**, *8*, 16.

[91] G. D. Fu, L. Q. Xu, F. Yao, K. Zhang, X. F. Wang, M. F. Zhu, S. Z. Nie, *ACS Appl. Mater. Interfaces* **2009**, *1*, 239–243.

[92] G.-D. Fu, L.-Q. Xu, F. Yao, G.-L. Li, E.-T. Kang, *ACS Appl. Mater. Interfaces* **2009**, *1*, 2424–2427.

[93] H. Yang, Q. Zhang, B. Lin, G. Fu, X. Zhang, L. Guo, *J. Polym. Sci. A Polym. Chem.* **2012**, *50*, 4182–4190.

[94] P. Viswanathan, E. Themistou, K. Ngamkham, G. C. Reilly, S. P. Armes, G. Battaglia, *Biomacromolecules* **2015**, *16*, 66–75.

[95] O. I. Kalaoglu-Altan, R. Sanyal, A. Sanyal, *Polym. Chem.* **2015**, *6*, 3372–3381.

[96] J. Wang, H. E. Naguib, A. Bazylak, N. C. Goulbourne, Z. Ounaies, *SPIE Proc* **2012**, *8342*, 83420F–1.

[97] G. Y. Han, B. Guo, L. W. Zhang, B. S. Yang, *Adv. Mater.* **2006**, *18*, 1709–1712.

[98] E. Chiellini, A. Corti, S. D'Antone, R. Solaro, *Prog. Polym. Sci.* **2003**, *28*, 963–1014.

[99] S. Hiessl, D. Böse, S. Oetermann, J. Eggers, J. Pietruszka, A. Steinbüchel, *Appl. Environ. Microbiol.* **2014**, *80*, 5231–5240.

[100] A. Linos, M. M. Berekaa, R. Reichelt, U. Keller, J. Schmitt, H.-C. Flemming, R. M. Kroppenstedt, A. Steinbuchel, *Appl. Environ. Microbiol.* **2000**, *66*, 1639–1645.

[101] U. Witt, R.-J. Müller, J. Augusta, H. Widdecke, W.-D. Deckwer, *Macromol. Chem. Phys.* **1994**, *195*, 793–802.

[102] A. M. Reed, D. K. Gilding, *Polymer* **1981**, *22*, 499–504.

List of abbreviations

AFM	atomic force microscope
3DF	3D fiber deposition
AAc	acrylic acid
AAPBA	3-acrylamidophenylboronic acid
ABP	4-acryloyloxybenzophenone
AIBN	azobisisobutyronitrile
Ala	Alanine (amino acid)
Am	acrylamide
AN	acrylonitrile
ASC	adipose-derived stem cells
ATR	attenuated total reflection
ATRP	atom transfer radical polymerization
B. mori	Bombyx mori (silkworm)
B. subtilis	Bacillus subtilis
BPA	bisphenol-A
BPAm	N-(4-benzoylphenyl) acrylamide
BPO	benzoylperoxide
BR	polybutadiene
BSE	back-scattered electrons
CAS	Chemical Abstracts Service
Ce	entanglement concentration
CHCl$_3$	chloroform
CL	ε-caprolactone
CR	polychloroprene
CVD	chemical vapor deposition
Đ	molar mass dispersity
DCM	dichloromethane
DIN	German Institute for Standardization
DLS	dynamic light scattering
DMF	dimethylformamide
DMSO	dimethyl sulfoxide
DSC	differential scanning calorimetry
DTG	differential thermogravimetry
E. coli	Escherichia coli (bacterium)
ECH	epichlorohydrin
ECM	extra cellular matrix
EDC	1-ethyl-3-(3-dimethyl-aminopropyl)-1-carbodiimide hydrochloride
EDMA	ethylene dimethacrylate
EDX	energy-dispersive X-ray spectroscopy
EtOH	ethanol
FDA	US food and drug administration
FT-IR	Fourier transform infrared spectroscopy
GC–MS	gas chromatograph–mass spectrometer
Gly	Glycine (amino acid)
GPC	gel permeation chromatography
HA	hydroxyapatite

HDPE	high-density polyethylene
HEMA	2-hydroxyethylmethacrylate
HEPΛ	high efficiency particulate air
HFIP	1,1,1,3,3,3–hexafluoro-2-propanol
HRTEM	high-resolution transmission electron microscope
HSP	Hansen solubility parameters
IR	infrared (spectroscopy)
IUPAC	International Union of Pure and Applied Chemistry
LBL	layer-by-layer
LCST	lower critical solution temperature
LDPE	low-density polyethylene
LED	light-emitting diode
LLDPE	linear-low-density polyethylene
MA	methyl acetate
MALDI-TOF	matrix-assisted laser desorption/ionization - time of flight
MDSC	modulated differential scanning calorimetry
MEK	methyl ethyl ketone
MeOH	methanol
MFC	microbial fuel cells
MIBK	methyl isobutyl ketone
M_n	number average molar mass
MPEG	α-Hydroxy-ω-methoxy-poly(ethylene glycol)
M_v	viscosity average molar mass
M_w	mass average molar mass
M_z	centrifugation average molar mass
NHS	N-hydroxysuccinimide
NMP	nitroxide mediate polymerization
NMR	nuclear magnetic resonance spectroscopy
ODA	4,4′-oxydianiline
P3HB	poly-(R)-3-hydroxybutyrate
P3HV	poly-3-hydroxyvalerate
PA	polyamide
PAA	poly(amic acid)
PAAc	poly(acrylic acid)
PAN	polyacrylonitrile
PBI	polybenzimidazole
PBT	poly(butylene terephthalate)
PC	polycarbonate
PCL	poly(ε-caprolactone)
PCL-b-MPEG	polycaprolactone-block-methoxypoly(ethylene glycol)
$Pd(OAC)_2$	palladium acetate
PE	polyethylene
PEG	poly(ethylene glycol)
PEGDA	poly(ethylene glycol) diacrylate
PEO	poly(ethylene oxide)
PEOT	poly(ethyleneoxide)-terephthalate
PET	poly(ethylene terephthalate)
PF	pyridinium formate
PHA	poly(hexamethylene adipate)

PI	polyimide
PI	cis-1,4-polyisoprene
PLA	polylactide
PLGA	poly(lactide-co-glycolide)
PLLA	poly(L-lactide)
PMDA	pyromellitic dianhydride
PMMA	poly(methyl methacrylate)
PNIPAm	poly(N-isopropylacrylamide)
POEGMA	poly[oligo(ethyleneglycol) methacrylate]
POM	poly(oxymethylene)
PP	polypropylene
PPO	poly(propylene oxide)
PPX	poly(p-xylylene)
PS	polystyrene
PTFE	polytetrafluoroethylene
PTT	poly(trimethylene terephthale)
PU	polyurethane
PVAc	poly(vinyl acetate)
PVB	poly(vinyl butyral)
PVC	poly(vinyl chloride)
PVD	physical vapor deposition
PVDF	poly(vinylidene fluoride)
PVDF-HFP	poly(vinylidenefluoride-hexafluoropropylene)
PVOH	poly(vinyl alcohol)
PVP	poly(vinyl pyrrolidone)
QCM	quartz crystal microbalance
RAFT	reversible addition-fragmentation chain – transfer polymerization
Ref.	reference
ROP	ring-opening polymerization
RP	rapid prototyping
SDS	sodium dodecyl sulfate
SE	secondary electrons
SEM	scanning electron microscope
Ser	Serine (amino acid)
SF	silk fibroin
SMC	smooth muscle cells
$Sn(oct)_2$	tin(II) 2-ethylhexanoate
TBAB	tetra-n-butylammonium bromide
Tcryst	crystallization temperature
TEBAC	benzyltriethylammonium chloride
TEM	transmission electron microscope
TFA	trifluoracetic acid
TFE	2,2,2-triflourothanol
TG	thermogravimetry
Tg	glass transition temperature
TGA	thermogravimetric analysis
TGS	thermodynamically good solvent
THF	tetrahydrofuran
Tm	melting temperature

TPE	thermoplastic elastomer
TPS	thermodynamically poor solvent
TPU	thermoplastic polyurethane
TUFT	tubes by fiber templates
UCST	upper critical solution temperature
UHMWPE	ultra-high-molecular-weight polyethylene
ULPA	ultra low penetration air
UV	ultraviolet light
UV-Vis	ultraviolet-visible spectrophotometry
VP	4-vinylpyridine
WAXS	wide-angle X-ray scattering
η_0	zero shear viscosity
μmPCL/Col	polycaprolactone-type I bovine collagen

Polymer index

Subject Index

Made in the USA
Lexington, KY
16 February 2018